Guidelines for the Design of Products Equipped with Phase Change Material

Guidelines for the Design of Products Equipped with Phase Change Material

By

Barbara Pause

Cambridge
Scholars
Publishing

Guidelines for the Design of Products Equipped with Phase Change Material

By Barbara Pause

This book first published 2024

Cambridge Scholars Publishing

Lady Stephenson Library, Newcastle upon Tyne, NE6 2PA, UK

British Library Cataloguing in Publication Data
A catalogue record for this book is available from the British Library

ISBN (10): 1-0364-0457-9
ISBN (13): 978-1-0364-0457-4

TABLE OF CONTENTS

CHAPTER ONE

INTRODUCTION

1.1 Market development and current applications

Phase change materials have been utilized as thermal storage means for quite some time. Currently, the demand for energy-efficient products, rising environmental consciousness, and laws addressing climate change drive growth in the market related to phase change materials. The global phase change material market was estimated to have a value of USD 1.78 billion in 2021 by the market research organization "Straits Research", based in India, and was forecasted by this organization to be about USD 7.60 billion by 2030 [1]. This means that the market would quadruple in size within about a decade. Other market research groups are more conservative in their predictions. The US-based market research and consulting company "Reports and Data" determined that the phase change material market had a value of USD 1.65 billion in 2022 and forecasted a market size of USD 3.62 billion by 2032 [2]. Under these considerations, the market would double in this decade. But even with this deviation in mind, there will be substantial growth in utilizing phase change materials in the coming years.

Currently, phase change materials are mainly used in:

- Building components, such as floor heating systems, space cooling installations, hot water storage, and thermal buffers for heat pumps.
- Cold chain and shipping components, i.e., transport containers and cold storage units.
- Electronics components comprising heat dissipation systems.

In these applications, phase change materials are utilized for thermal storage purposes where heat and cold storage are provided. Another objective for the current application of phase change material is heat dissipation. Heat is drawn away from a source, such as an electronic device/component, and stored in a heat sink.

Related to textiles, the major applications of phase change materials are currently in bedding products and cooling vests [3, 4].

1.2 Objectives of the Design Guidelines

Besides the applications in building components, cold chain and shipping components, and electronics cooling solutions, there are many more areas where products could benefit from phase change material technology. There is also a wide range of possible applications in which textiles could be used as the carrier for phase change material.

The efficiency of thermal storage and heat dissipation is always limited by the amount of phase change material utilized in the system. Furthermore, there is a challenge regarding recharging the phase material used in such a system. If there is only a partial recharge, the phase change material will perform at a limited capacity in the next cycle of operation.

In contrast, applications in which the phase change material's heat absorption and heat emission features are used alternately and in equal proportion have a greater potential for providing long-lasting thermal effects. These systems act, for instance, as thermostats keeping enclosed systems at a nearly constant temperature. Examples of such applications will be discussed in the book.

Another feature that is obtained when applying phase change material to a sheet-like substrate is the potential of a temperature-controlled heat flux through it. For instance, many applications in heat protection would benefit from this feature. Because these changes in the heat flux are triggered by an outside stimulus, in this case, the temperature, substrates with the phase change material treatment are called "smart materials". The heat flux control feature is not utilized in current applications of phase change materials. However, examples of possible applications are included in the book and will be discussed.

Many years of experience developing products utilizing phase change material technology have demonstrated that some requirements must be considered in the design process. If these requirements were not met in the past, the products mostly failed consumer expectations regarding thermal performance and were often taken off the market in a short period. To avoid such pitfalls, the book will give designers guidelines and a toolbox to support successful product development. Furthermore, the book provides information about testing procedures applied to validate the thermal performance of the developed products.

References

1. Straits Research. 2023. “Phase Change Materials Market: Information by Type (Organic PCM, Inorganic PCM), Form (Encapsulated, Non-encapsulated) Application (Packaging, HVAC), and Region-Forecast till 2030”. https://straitsresearch.com.
2. Reports and Data. 2022. “Phase Change Material Market By Type (Organic, Inorganic, Bio-based) By Application (Building & Construction, HVAC, Energy Storage, Cold Chain & Packaging, Textile, Electronics, and Others), and By Region Forecast to 2032”. https://www.reportsanddata.com.
3. MarketWatch. 2023.” 2030 Phase Change Material (PCM) Market Research”. Press Release on June 19, 2023, https://www.marketwatch.com
4. Brütting, Michael, Haussmann, Thomas, Mehling, Harald. 2022. “PCM products and their fields of application – An overview of the state in 2020/2021”. Journal of Energy Storage, 51. https://doi.org/10.1016/j.est.2022.104354

CHAPTER TWO

PHASE CHANGE TECHNOLOGY

2.1 Phase change effect

The most well-known thermal energy storage occurs when a material, such as water, is heated in an ordinary heating process. The amount of sensible heat absorbed in such a heating process depends on the weight of the heated material, the temperature difference covered in the process, and the material-related specific heat capacity. However, because of the low energy density of such systems, sensible heat storage requires large volumes to be effective.

A more efficient way to store thermal energy is provided by the application of phase change material. Phase change materials possess the ability to change their physical state from solid to liquid and vice versa within certain temperature ranges. During the phase change from a solid into a liquid, a change from a defined oriented solid crystalline structure into a randomly oriented liquid state occurs. This is due to an increase in vibrational energy generated by heat which breaks the bonds between the molecules. The energy necessary to cause the phase change material to change from a solid into a liquid, i.e., to break down the bonding responsible for the solid structure, is absorbed from the surrounding area during a heating process. The result is that the temperature increase, which usually occurs in a material while heated, is interrupted when the melting temperature of the phase change material is reached. During the entire melting process, the temperature of the phase change material remains constant. When the phase change process is complete and the material is further heated, its temperature will continue to rise (Fig. 2-1).

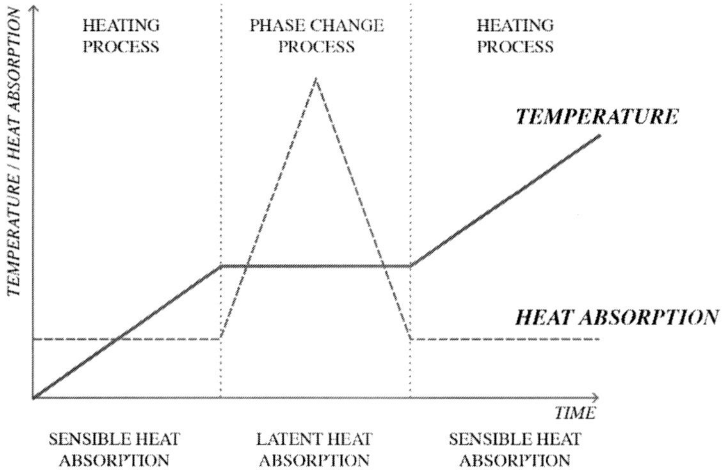

Fig. 2-1: Phase change effect.

The latent heat is stored in the phase change material and released in a reverse cooling process when the crystallization temperature of the phase change material is reached. A nucleation process then rearranges the molecules into the crystalline lattice. The material temperature also remains constant during the reverse phase change from a liquid into a solid. However, when the phase change is complete, a continued cooling process results in a further temperature decrease.

Water shall be used to compare the amount of latent heat absorbed by a phase change material during the actual phase change with the amount of sensible heat absorbed in an ordinary heating process. If ice melts into water, it absorbs a latent heat of about 335 Joules per gram. If water is further heated, only 4 Joules per gram of sensible heat is taken up while the temperature rises by one degree. Therefore, the latent heat absorption during the phase change from ice into water is nearly 100 times higher than the sensible heat absorption during the heating process of water outside the phase change temperature range. In other words, water needs to be heated from about 1 °C up to 85 °C to absorb the same amount of heat that is absorbed during the melting process of ice.

In addition to the variances in the amount of heat that can be absorbed or released, there are several other differences between the sensible heat exchange and the latent heat exchange, which are summarized in Table 2-1.

Characteristics	Sensible heat absorption/release	Latent heat absorption/release
Heat exchange process	continuously	discontinuously
Material temperature	increases/decreases	remains constant
Material density	remains constant	increases/decreases

Table 2-1: Comparison between sensible heat exchange and latent heat exchange.

The major advantage of the latent heat exchange compared to the sensible heat exchange is the much higher amount of heat that can be absorbed or released without a significant change in the material's temperature. On the other hand, a disadvantage in using a latent heat storage material or phase change material is its density change resulting from the phase transition, leading to a change in the material's volume. But altogether, the materials that store latent heat are thermally more effective than those that store sensible heat on both the weight and volume bases.

Besides the solid-liquid phase changes, solid-solid phase changes have also been experienced where the crystalline structure is only changed into a semi-crystalline or amorphous structure when a phase transition at a given temperature in a heating process occurs. During this solid-solid phase transition, the material absorbs latent heat. The latent heat is released in a reverse cooling process at the phase transition temperature. At this point, the semi-crystalline or amorphous structure transforms back into the defined crystalline structure. Unlike phase change materials with solid-liquid transitions, solid-solid phase change materials mainly retain their physical properties. Their density changes only slightly, and their volume remains nearly constant. These properties create advantages for a variety of applications.

In addition to the solid-liquid and the solid-solid phase change, there are other possible transitions in the physical state of a material, such as

- Solid-vapor transitions.
- Liquid-vapor transitions.

Solid-vapor and liquid-vapor transitions are related to a much higher amount of heat to be exchanged compared to solid-liquid transitions as well as solid-solid transitions. For instance, the amount of heat absorbed during the vaporization of water is about 10 times higher than the amount of heat absorbed during the melting process from ice into water. However, solid-vapor or liquid-vapor transitions cannot be practically used in many applications because of their tremendous volume, density, and pressure changes resulting from the vaporization process. Furthermore, the reverse transition process from vapor into solids or liquids is technically impossible for many applications.

2.2 Phase change materials

2.2.1 Overview and selection requirements

Besides water, more than 500 natural or synthetic phase change materials are known which differ from one another by the temperatures at which their phase changes take place and their latent heat storage capacities [1]. Phase change materials should be selected for a specific application, considering the appropriate phase change temperature range and the necessary latent heat storage capacity.

Furthermore, the phase change material selection shall also be predicated on physical and chemical properties desirable or undesirable for the prevailing application. For many applications, the selected phase change material must be nontoxic, noncorrosive, non-hygroscopic, and non-flammable.

Further requirements are, for instance, low volume changes during the phase transition, low vapor pressures near the melting points to prevent excessive stress on surrounding storage containers, and high thermal conductivity to avoid the appearance of thermal gradients.

In addition, the selected phase change material should be stable to repeated thermocycles. Furthermore, it must melt and crystallize in a congruent manner. The phase change material should not show a supercooling behavior where a liquid phase change material is cooled below the phase change temperature range without forming a solid phase. Finally, price and availability also play an important role in selecting a phase change material for a specific application.

Phase change materials are divided into the categories summarized in Fig. 2-2. Phase change materials transferring from a solid into a liquid and vice versa are categorized as organic phase change materials and inorganic phase change materials. Organic phase change materials comprise paraffin and non-paraffinic organics such as fatty acids and polyethylene glycols. Salt hydrates, salt eutectics, and metallics are summarized under the term 'inorganic phase change material'. Solid-solid phase change materials include organics, organometallics, inorganics, and polymerics.

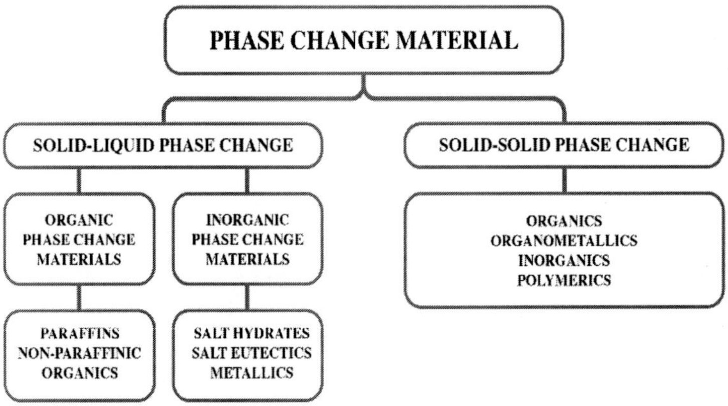

Fig. 2-2: Phase change material categories.

2.2.2 Organic phase change materials

Paraffin

The paraffins of type C_nH_{2n+2} are parts of the family of saturated hydrocarbons and possess similar properties. The materials with carbon atom numbers below pentane (C_5H_{12}) are gases at room temperature.

From pentane (C_5H_{12}) to pentadecane ($C_{15}H_{32}$), paraffins are liquids at room temperature. Paraffins with higher carbon atom numbers are waxy solids when stored at room temperature.

The melting temperature of the paraffins increases with the number of carbon atoms in the hydrocarbon chain, as shown in Fig. 2-3.

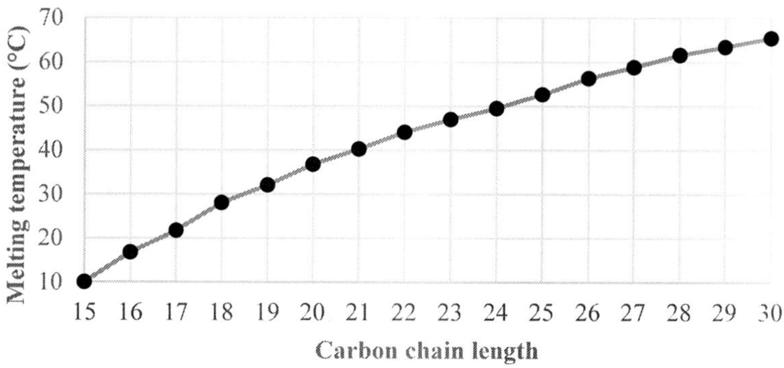

Fig. 2-3: Melting temperatures of different paraffins depending on their carbon chain length.

Paraffins possess relatively high latent heat storage capacities [2]. Some of them are summarized in Table 2-2. Table 2-2 also contains the melting and crystallization temperatures of the selected paraffins.

Paraffin	Formula	Melting temperature, °C	Crystallization temperature, °C	Latent heat storage capacity, J/g
Hexadecane	$C_{16}H_{34}$	16.7	12.2	237
Heptadecane	$C_{17}H_{36}$	21.7	16.5	213
Octadecane	$C_{18}H_{38}$	28.0	25.4	244
Nonadecane	$C_{19}H_{40}$	32.0	26.4	222
Eicosane	$C_{20}H_{42}$	36.7	30.6	247
Hen-eicosane	$C_{21}H_{44}$	40.2	35.9	213

Table 2-2: Phase change temperatures and latent heat storage capacities of selected paraffins.

A comparison of the latent heat storage capacities summarized in Table 2-2 shows that the paraffins with even numbers of carbon atoms possess slightly higher latent heat storage capacities than the paraffins with odd numbers of carbon atoms. The distance between melting and crystallization temperature is less than 10 degrees Celsius for all paraffins. Thus, the paraffins appear to be self-nucleating.

The "pure" paraffins are nontoxic and noncorrosive. They are non-hygroscopic and absorb little or no moisture at high humidities or when exposed to liquid water.

Paraffins show congruent melting and freezing behavior, so they are stable regarding repeated thermocycles. However, prolonged exposure to temperatures above 150°C can chemically degrade them.

A sample of an eicosane paraffin at room temperature is shown in Fig. 2-4.

Fig. 2-4: Eicosane paraffin piece at room temperature.

Besides all these advantages, there are also some disadvantages related to the use of paraffin. Paraffins are known for extremely low thermal conductivities. They have insulating properties comparable to the best insulators. However, the thermal conductivity can be enhanced, for instance, by mixing them with metallic fillers [2].

The density of the paraffins decreases when they melt and become liquid. These density changes lead to an average volume expansion of about 8 percent. Densities and thermal conductivities in the solid states and liquid states of the paraffins included in Table 2-2 are summarized in Table 2-3.

Furthermore, like other organic compounds, paraffins will burn if ignited. Paraffin can be incorporated or encapsulated in non-flammable materials to avoid such an ignition. In addition to that, paraffin could be made self-extinguished by formulating them with small quantities of specific fire-retardant additives such as organic halides or antimony oxides. Tests have shown that there is no significant change in the melting and crystallization behavior of the paraffins after combining them with fire-retardant additives.

Paraffin	Density (liquid), kg/m^3	Density (solid), kg/m^3	Thermal conductivity (liquid), W/m K	Thermal conductivity (solid), W/m K
Hexadecane	770	835	0.150	0.183
Heptadecane	774	839	0.152	0.200
Octadecane	777	844	0.155	0.218
Nonadecane	783	848	0.161	0.224
Eicosane	789	854	0.172	0.236
Hen-eicosane	792	859	0.175	0.243

Table 2-3: Densities and thermal conductivities of selected paraffins in liquid states and solid states.

There are two ways of manufacturing paraffins:

- Ethylene polymerization.
- Fractionation of petroleum refining residues.

Paraffin products resulting from the polymerization of ethylene monomers possess high purities. They are marketed for a comparatively high price.

Paraffins made as byproducts of petroleum refining are sold at much lower prices in tank car quantities. These paraffins are technical grades with a purity of about 95 percent or blends of paraffins with different chain lengths.

In textile applications, technical grade paraffin or paraffin blends are often used. Compared to pure-grade paraffin, they usually possess lower heat storage capacities and melt as well as freeze over a wider temperature range. Fig. 2-5 shows a comparison of melting temperature ranges of pure octadecane, technical grade octadecane, and a blend of 40 percent octadecane (C_{18}) with 38 percent heptadecane (C_{17}) and smaller portions of nonadecane (C_{19}), and eicosane (C_{20}) paraffin.

Technical grade octadecane possesses a slightly lower melting temperature range than pure octadecane, which needs to be taken into consideration for practical use of technical grade octadecane.

Blending octadecane mainly with heptadecane leads to a further decrease in the melting temperature range. A comparison of the melting temperature ranges of the different octadecane variants is shown in Fig. 2-5.

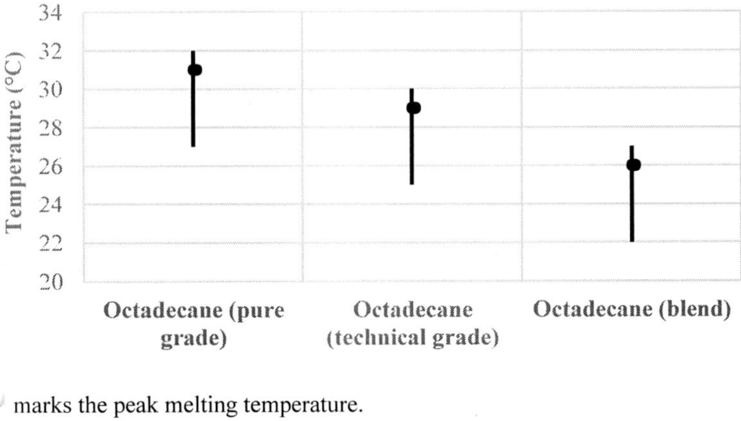

Fig. 2-5: Comparison of the melting temperature ranges of different variants of octadecane paraffin.

Reducing the purity of octadecane lowers the temperature range in which the melting process takes place. In addition, the reduction in the purity of octadecane leads to an extension of the crystallization temperature range, as shown in Fig. 2-6.

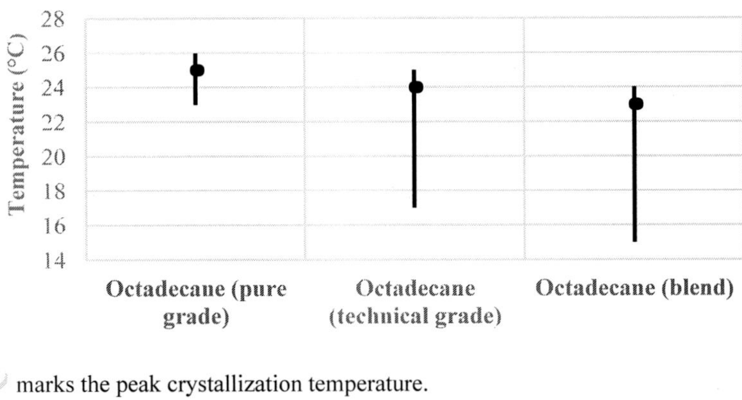

Fig. 2-6: Comparison of the crystallization temperature ranges of different variants of octadecane paraffin.

Finally, Fig. 2-7 summarizes the latent heat storage capacities of the different variants of the octadecane paraffine. The results shown in Fig. 2-7 indicate that the latent heat storage capacity of technical grade octadecane is about 13 percent lower than the latent heat storage capacity of pure octadecane. The investigated blend possesses a latent heat storage capacity which is about 35 percent lower than that of the pure octadecane. Similar results have been received for other paraffin configurations.

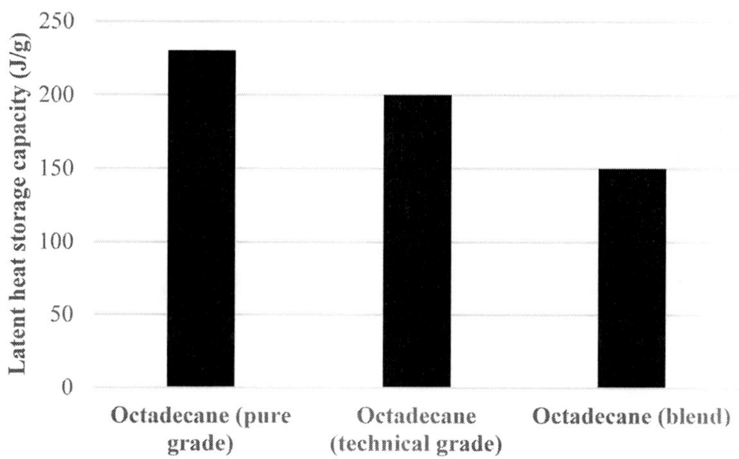

Fig. 2-7: Comparison of the latent heat storage capacities of different variants of octadecane paraffin.

Non-paraffinic organics

Non-paraffinic organics include a wide variety of organic materials such as fatty acids, esters, alcohols, and glycols. Each of these materials has its own characteristics, unlike paraffins which possess very similar properties.

In terms of potential applications, it needs to be considered that some of the non-paraffinic organics are flammable. Many of them have low flash points, and they will decompose at elevated temperatures. They possess low thermal conductivities like those of the paraffins. To improve the heat transfer to and from them, the use of metallic fillers is recommended.

Some of the non-paraffinic organics, i.e., many of the fatty acids are plant-based or bio-based products. Therefore, they are considered as "green" materials. A selection of fatty acids is summarized in Table 2-4.

Fatty acid	Formula	Melting temperature, °C	Latent heat storage capacity, J/g
Acetic acid	CH_3COOH	17	184
Elaidic acid	$C_8H_{17}C_9H_{16}COOH$	47	218
Myristic acid	$CH_3(CH_2)_{12}COOH$	58	199
Stearic acid	$CH_3(CH_2)_{16}COOH$	69	199

Table 2-4: Melting temperatures and latent heat storage capacities of selected fatty acids.

Fatty acids exhibit excellent melting and crystallization characteristics [3]. Their latent heat storage capacities are comparable to those of paraffins. A myristic acid sample at room temperature is shown in Fig. 2-8.

Fig. 2-8: Myristic acid sample at room temperature.

Both paraffins and fatty acids possess a white, waxy, and soft appearance in their solid states. While paraffins are often sold in chunks such as seen in Fig. 2-4, fatty acids are submitted in the form of flakes.

Polyethylene glycol is a polyether compound that is derived from petroleum. Polyethylene glycols with molar masses between 400 g/mol and 20,000 g/mol possess melting points between 0 °C and 70 °C. Their latent heat storage capacities increase with a rise in the molar mass and range from about 90 J/g to 190 J/g. In comparison of the latent heat storage capacities of the paraffin nonadecane and polyethylene glycol 1000 operating in the same temperature range, the latent heat storage capacity of polyethylene glycol 1000 is only about 65 percent of the latent heat storage capacity of nonadecane.

Polyethylene glycols are chemically and thermally stable, nontoxic, noncorrosive, and biodegradable. However, a substantial disadvantage is the large difference between their melting and freezing temperatures which leads to incongruent melting behavior, resulting in a significant loss in latent heat storage capacity even when used only over a short period [4].

Applications of organic phase change material

Because of the temperature range in which some organic phase change materials, such as paraffin and fatty acids, operate and their relatively high latent heat storage capacities, they are especially suitable for applications in garments, sporting goods, and medical equipment. Furthermore, their performance does not change over time even under repeated thermocycles. By mixing different paraffins, specific temperature ranges can be covered, and the desired application temperature range can be met.

2.2.3 Inorganic phase change materials

Salt hydrates

Salt hydrates are alloys of inorganic salts and water. The water molecules are absorbed into the crystalline structure of the salts. Salt hydrates are non-combustible and, therefore, meet the fire-resistance requirements of a variety of products and applications.

The most attractive properties of salt hydrates are their comparatively high latent heat storage capacities, high thermal conductivities, and small volume changes at melting. However, most salt hydrates are corrosive which can be problematic when storing them in metal containers. Salt hydrates possess higher densities than organic phase change materials leading to higher masses per volume, i.e., higher weights.

Salt hydrates show three kinds of melting behavior: congruent, semi-congruent, and incongruent melting. Incongruent and semi-congruent melting salt hydrates are inexpensive, but they are difficult to handle [5]. Their components get separated during the melting process. In a reverse cooling process, the components are unable to recombine appropriately to form the original crystalline structure. The lack of reversible melting and freezing makes the salt hydrates with incongruent and semi-congruent melting and freezing behavior unsuitable for many applications. On the other hand, salt hydrates melting congruently are expensive [6]. The thermal properties of three salt hydrates are given in Table 2-5.

Salt hydrate	Formula	Melting temperature, °C	Latent heat storage capacity, J/g
Lithium nitrate trihydrate	$LiNO_3$ x 3 H_2O	30	296
Sodium hydrogen phosphate dodecahydrate	Na_2HPO_4 x 12 H_2O	36	280
Sodium acetate trihydrate	$NaC_2H_3O_2$ x 3 H_2O	58	265

Table 2-5: Melting temperatures and latent heat storage capacities of selected salt hydrates.

A sample of the salt hydrate sodium acetate trihydrate is shown in Fig. 2-9. The crystalline structure of the salt hydrate particles can be clearly identified in the picture.

Fig. 2-9: Sodium acetate trihydrate sample at room temperature.

The incongruent melting and freezing behavior of various salt hydrates is especially experienced when the materials do not freeze at their solidification point, requiring substantial subcooling before solidification nucleation begins. This problem can be solved by using nucleation agents [7]. These agents provide a surface from which the formation of a crystalline structure can occur faster, preventing what is called the supercooling effect. The elimination of the supercooling in these salt hydrates ensures that their latent heat storage capacity does not decrease over time, i.e., it remains stable over many thermocycles.

Salt eutectics

Salt eutectics are created by mixing two or more salt hydrates. They melt and crystallize at specific temperatures, often deviating from those of the single components. Their behavior is analogous to congruent melting salt hydrates. For instance, a stable eutectic mixture consists of 49.3 percent Mg Cl x 6 H_2O and 50.7 percent Mg $(NO_3)_2$ x 6 H_2O, melting at 58 °C and absorbing a latent heat of about 132 J/g.

Some of the salt eutectics possess not one but several melting temperatures. These melting temperatures can be changed by altering the ratio of the eutectic components or the water content.

Salt eutectics have advantages in high thermal conductivity and large energy density. A disadvantage is their corrosivity.

Applications of inorganic phase change materials

Salt hydrates and salt eutectics are preferably used for industrial applications where fire-resistant requirements must be met. Salt hydrates can be mixed to create salt eutectics that cover the desired temperature ranges where the phase change occurs. The high thermal conductivity of salt hydrates ensures sufficient heat transfer to and from the phase change material.

Metallics

Metallics are very heavy compared to other phase change materials, and their latent heat storage capacities per unit weight are comparatively low. But compared based on a volume base, their latent heat storage capacities are acceptable. The advantages of metallics are their high thermal conductivity and sufficient thermal stability. An example of a metallic phase change material is gallium which melts at a temperature of 30 °C and possesses a latent heat storage capacity of about 80 J/g. The thermal conductivity of gallium is about 34 W/m K. An interesting feature is the shrinkage of the material during melting [1]. The volume change is about 3 percent. Gallium is nontoxic but very corrosive. Compared to other phase change materials, gallium is also very expensive and, therefore, not suitable even for industrial applications.

2.2.4 Solid-solid phase change materials

Phase change materials that transfer from one solid material into another solid material retain their bulk solid properties within the temperature range of the phase change. By staying solid, leakage problems experienced at solid-liquid phase changes can be avoided. They can be blended into a matrix material without a need for encapsulation. Solid-solid phase change materials show only small volume changes by undergoing a phase transition [8].

Organics

Organic solid-solid phase change materials have phase transitions in a temperature range between about 60 °C and 190 °C. Their latent heat storage capacities range from 180 J/g to 300 J/g. Some examples of organic solid-solid organic phase change materials are summarized in Table 2-6.

Organic solid-solid phase change material	Formula	Melting temperature, °C	Latent heat storage capacity, J/g
Tris hydroxymethyl-ethane	$C_5H_{12}O_3$	82	190
Tris hydroxymethyl-aminomethane	$C_4H_{11}NO_3$	131	285
Pentaerythritol	$C(CH_2OH)_4$	185	300

Table 2-6: Melting temperatures and latent heat storage capacities of selected organic solid-solid phase change materials.

The thermal conductivity of these organic solid-solid phase change materials is only low and comparable to the thermal conductivity of paraffins and fatty acids. For instance, the thermal conductivity of pentaerythritol is about 0.2 W/m K. The appearance of pentaerythritol is a fine white powder shown in Fig. 2-10.

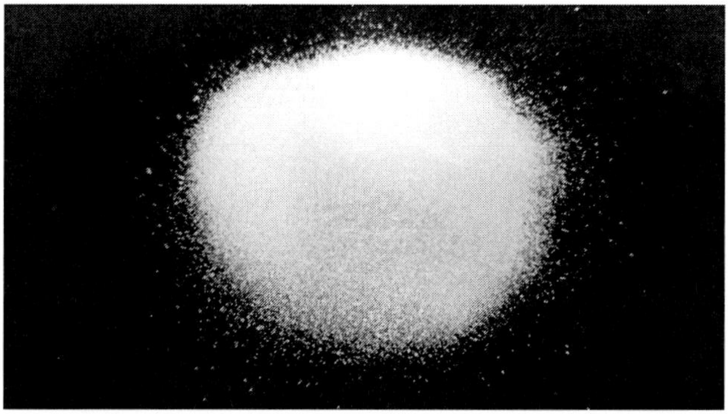

Fig. 2-10: Pentaerythritol sample.

Organometallics

Organometallics consist of alternate inorganic and organic layers. They create a sandwich-like crystalline structure.

The phase transition is only performed in the organic part. Organometallics consist of a paraffinic chain with alkyl-ammonium groups creating the organic part and a metal or a halogen building the inorganic part. Their solid-solid transition temperatures are between 20 °C and 90 °C. The latent heat storage capacities of these phase change materials range from 60 J/g to 150 J/g. Since the inorganic components (metal and halogen) of the organometallics do not participate in the phase transition, they contribute only to the weights of the phase change materials and not their latent heat storage capacities.

This explains the low numbers obtained for the latent heat storage capacities of these solid-solid phase change materials. Organometallics show a very small volume change during the phase transition. Due to the metal components in the materials, their thermal conductivities are very high.

Inorganics

Inorganic solid-solid phase change materials operate between 680 °C and 900 °C. They have very low latent heat storage capacities, around 50 J/g. An example of an inorganic solid-solid phase change material is an iron-cobalt alloy. Inorganic solid-solid phase change materials possess high thermal conductivities between 0.3 W/m K and 5 W/m K [8].

Polymerics

Polymeric solid-solid phase change materials consist of hard and soft segments, and the phase change takes place only in the soft segments. For instance, polyethylene glycol serves as the soft segment in a polymeric phase change material, while polystyrene creates the hard segment functioning as the "backbone".

Polymeric solid-solid phase change materials operate in a temperature range between 10 °C and 65 °C and possess latent heat storage capacities varying in a wide range between 10 J/g and 200 J/g. The volume change during the phase transition is only small. They are thermally stable. The thermal conductivity of polymeric solid-solid phase change materials is with 0.1W/m K to 0.2 W/m K only low, i.e., in the range of the thermal conductivity of paraffin and fatty acids [8]. They are nontoxic but possess a low fire resistance.

Applications of solid-solid phase change materials

Based on the temperature range in which the inorganic solid-solid phase change materials are operating and their low latent heat storage capacities, they are not suitable candidates for current and probably also future applications. The polymeric and organometallic solid-solid phase change materials possess phase transitions in a more suitable temperature range, but their latent heat storage capacity is not high enough to be considered in current applications. In contrast, organic solid-solid phase change materials possess desirable latent heat storage capacities for a variety of industrial applications. The only disadvantage in utilizing these materials is their low thermal conductivity comparable to the thermal conductivity of paraffin and fatty acids.

References

1. Hale, D. V., Hoover, M. J., O'Neill, M. J. 1971. "Phase change materials handbook". NASA Contractor Report, Contract NAS8-25183, NASA Marshall Space Flight Center, Alabama, USA
2. Fleischer, Amy S. 2015. "Thermal Energy Storage Using Phase Change Materials: Fundamentals and Applications". Springer Briefs in Applied Science and Technology, Thermal Engineering and Applied Science, 2015 Edition, Springer Cham, Heidelberg, New York, Dordrecht, London

3. Chowdhury, Hasnat Noor, Hamja, Abu, Rahman, Mahbubur. 2013. "Phase Change Materials: Characteristics and Encapsulation". International Conference on Mechanical Engineering and Renewable Energy, Chittagong, Bangladesh, December 24-27, 2013
4. Onder, Emel, Sarier, Nihal. 2015. "Thermal regulation finishes for textiles". In: Functional finishes for textiles. 17-98. Woodhead Publishing Ltd. https://doi.org/10.1533/9780857098450.1.17
5. Buddhi, D., Chen, C.R., Sharma, Atul, Tyagi, V.V. 2009. "Review on thermal energy storage with phase change materials and applications". Elsevier, Renewable and Sustainable Energy Reviews. No. 13: 318-345.
https://doi.org/10.1016/j.rser.2007.10.005
6. Cabeza, Luisa F., Marin, Jose M., Mehling, Harald, Zalba, Belen. 2003. "Review on thermal energy storage with phase change: materials, heat transfer analysis and applications". Pergamon, Applied Thermal Engineering. No. 23: 251-283
7. Gluesenkamp, Kyle R., Graham, Samuel, Hirschey, Jason, Kumar, Navin, et al. 2021. Review of Low-Cost Organic and Inorganic Phase Change Materials with Phase Change Temperature between 0°C and 65°C". 6th International High Performance Buildings Conference at Purdue, USA, May 24-28, 2021
8. Fallahi, Ali, Granados-Focil, Sergio, Guldentops, Gert, Tao, Mingjiang, et al. 2017. "Review on solid-solid phase change materials for thermal energy storage: molecular structure and thermal properties". Elsevier, Applied Thermal Engineering. No. 127: 1427-1441.
https://doi.org/10.1016/j.applthermaleng.2017.08.161

CHAPTER THREE

ENCAPSULATION

3.1. Encapsulation requirements

The solid-liquid phase change materials must be properly contained to prevent dissolution out of an end-use product configuration while in their liquid stage. The encapsulation needs to fulfill several requirements:

- Accommodates a volume change related to the phase change from solid to liquid and vice versa.
- Ensures a sufficient heat transfer to and from the encapsulated phase change material.
- Is thermally stable.
- Possesses sufficient mechanical strength.
- Does not react chemically with the encapsulated phase change material.
- Does not change the phase itself.

There are four forms of encapsulation currently utilized or under investigation:

- Macro-encapsulation.
- Micro-encapsulation.
- Nano-encapsulation.
- Shape-stabilization-encapsulation.

3.2 Macro-encapsulation

Several companies use macroscale container systems to encapsulate phase change materials. Most of these containers are made of plastic, such as HDPD, or metal [1, 2]. Some of the containers possess a rectangular shape. These containers are manufactured in various sizes, as shown in Fig. 3-1.

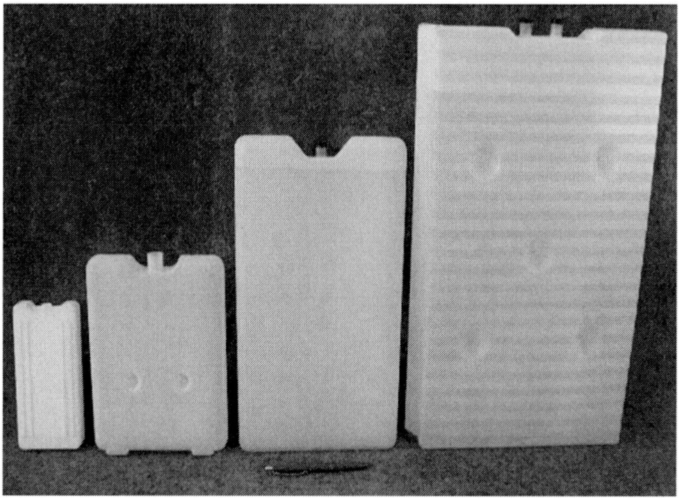

Fig. 3-1: Macroscale containers (source: Rubitherm GmbH, Germany).

In addition to these rectangular containers, there are plastic containers in other shapes like tubes, sticks, or balls. Furthermore, macro-encapsulated phase change materials are supplied in flexible, transparent polymer bags or flexible bags made of aluminum foils (Fig. 3-2).

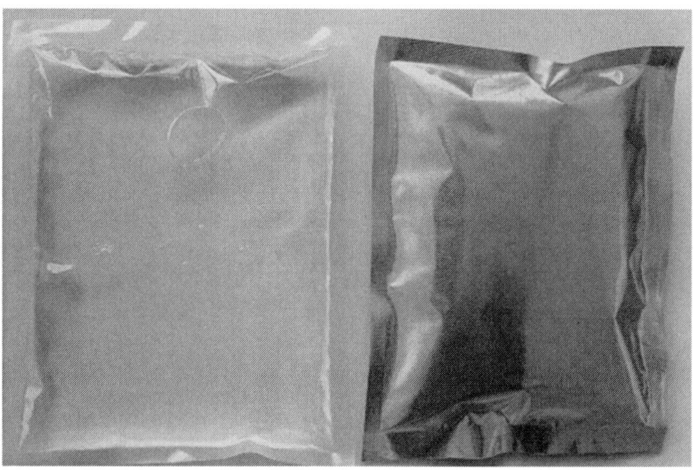

Fig. 3-2: Flexible bags filled with phase change material (source: Rubitherm GmbH, Germany).

The phase change materials are filled into these containers in their liquid state taking up the entire interior space of the container. When the phase change material later solidifies, there will be an air gap in the container due to the density increase and a subsequent volume decrease of the phase change material.

Because of the air's substantially lower thermal conductivity in comparison to the thermal conductivity of the phase change material, this will influence the heat transfer properties. For instance, the heat flow into the container will be reduced leading to a delay in the heat absorption by the phase change material.

Plastic containers should not be filled with phase change material possessing a high melting temperature. Filling the plastic container with a hot liquid will soften the plastic material which, in turn, can result in leaks. Metal containers are not suitable for salt hydrates due to their corrosivity. Cylindrical or spheric containers should be preferred in comparison to rectangular containers in terms of minimizing leak points.

In cases where structural strength is not an issue, thin-film plastic or aluminum bags could be used to encapsulate phase change material on a macro scale. The thin films are flexible enough to accommodate the volume change. However, it needs to be considered that thin-film bags can leak due to rupture or abrasion. The plastic material also ages, reducing the bag's flexibility.

The macroscale containment systems can be combined with various thermal conductivity enhancement methods necessary for specific applications [1]. Chapter 7 will discuss these methods.

3.3 Micro-encapsulation

To go away from the bulkiness of the macroscale container systems and to open the phase change material technology to new applications specifically in textiles or foams, tiny microcapsules containing the phase change material were developed. Besides the prevention of leakage of the melted phase change material out of a carrier material, micro-encapsulation increases the heat transfer rate to and from the phase change material, reduces the reactivity of the phase change material with the outside environment, and controls the change in volume as phase change occurs [2].

Micro-encapsulation is a process in which particles or individual droplets of phase change materials are coated with a continuous film to produce capsules with diameters in the size of micrometers. Phase change material microcapsules consist of two parts: the phase change material core and the shell material [1, 3, 4].

Microcapsules may be regular, i.e., spherical or irregular, depending mainly on the core phase change material and the formation of the shell material during production.

Currently, organic phase change materials such as paraffins and fatty acids are primarily microencapsulated. In lab trials, salt hydrates have also been microencapsulated. However, these microcapsules are not commercially available.

Shell materials

The shell can consist of an organic material such as acrylic resin or melamine formaldehyde resin, an inorganic material such as silica, calcium carbonate, or zinc oxide, and an organic-inorganic hybrid material such as polymethylmethacrylate (PMMA)-silica compound [3].

Properties of the various shell materials

Organic shells possess good sealing properties, good structural flexibility, and excellent resistance to the volume change happening under repeated phase transitions of the phase change material [2, 3]. Specifically, melamine formaldehyde resin shells are thermally stable and chemically compatible. Furthermore, melamine possesses flame-retardant properties that enhance the prevailing microcapsule's fire resistance.

Among the acrylic resins, the shells made of methacrylate copolymers, such as polymethylmethacrylate (PMMA) shells, are nontoxic, thermally stable, and chemically resistant. They are also easy to manufacture and relatively inexpensive. Organic shells are mainly used to encapsulate paraffines.

Compared to organic shell materials, inorganic shells are more rigid, have increased mechanical strength, and have higher thermal conductivity. Most of them are fire-resistant. Silica is a favorite inorganic shell material. Microcapsules made of silica are easy to produce. Inorganic shell materials are utilized to encapsulate paraffins, salt hydrates, and fatty acids [3].

Organic-inorganic shell materials combine the advantages of inorganic materials and organic materials. They possess increased rigidity, high thermal stability, and enhanced thermal conductivity due to the inorganic component, but at the same time, structural flexibility due to the organic part [3].

Manufacturing methods

Microcapsules containing phase change material can be manufactured in several ways. There are physical methods, chemical methods, and combinations of both [3, 4, 6]. The chemical methods are the most common methods for microencapsulating phase change materials.

Physical methods

There are physical methods in which the shells of the microcapsules are formed in a physical process such as drying, dehydration, or adhesion. The only physical method utilized for microencapsulating phase change materials is spray drying.

Spray drying is a low-cost commercial process. The encapsulation by spray drying is carried out by co-dissolving the phase change material core and the shell material in a solvent and spraying the mixture into a heated chamber where a fast evaporation of the solvent occurs. Specifically, the process comprises the preparation of an oil-water emulsion of the selected phase change material and the selected shell material. The solution is then atomized, and the mist is sprayed into the heated chamber, where hot air is used to convert the mist into powder. Finally, the microcapsules are filtered out [3, 5, 6].

Chemical methods

The chemical methods utilized to create microcapsules of phase change materials are all based on polymerization. Polymerization techniques comprise emulsion, suspension, and interfacial polymerization.

In all three polymerization techniques, a chemical reaction is initiated to generate a fast growth of a thin polymeric coating onto the surface around droplets consisting of the liquid phase change material.

For instance, when carrying out the emulsion polymerization, small droplets of the liquid phase change material are dispersed into a carrier fluid that already contains the shell material of the microcapsules. Then, an emulsion stabilizer is also added to the carrier fluid. The phase change material is encapsulated by the formation of a compact polymer layer around it while the carrier fluid evaporates. Polymethylmethacrylate (PMMA) is commonly used as shell material in the formation of microcapsules utilizing this method [1, 3].

An oil-water emulsion is created in the first step to perform the interfacial polymerization, which contains the phase change material and a hydrophobic monomer. Then, a hydrophilic monomer is added to the system to initiate polymerization. Finally, the microcapsules are obtained through filtering, washing, and drying. Melamine/formaldehyde is a shell material that is commonly used in interfacial polymerization. [3].

Physical-chemical methods

A physical-chemical method is a combination of physical and chemical procedures.

Physical processes such as phase separation are combined with chemical processes such as crosslinking or condensation to obtain microcapsules containing phase change material. One physical-chemical method is called coacervation. Coacervation occurs when two polymers whose macromolecules bear opposite charges interact with each other. This happens after an emulsion is prepared by dispersing the phase change material into an aqueous polymer solution. A second aqueous polymer solution with opposite charges is added. This leads to the deposition of the shell material on the surface of the phase change material droplets by electrostatic attraction. Cross-linking or thermal treatment is finally carried out to harden the microcapsules.

Another physical-chemical microencapsulation method is the sol-gel procedure. In this method, the reactive materials, including the phase change material, the emulsifier, and the solvent, are dispersed in a continuous phase to create a colloidal solution. Through condensation polymerization, a gel system with a three-dimensional structure is formed from which the microcapsules are created in a curing and drying procedure. Silica is used as a shell material when microcapsules are produced utilizing this method [3, 4].

Properties of the microcapsules containing phase change materials

Particle size

The particle size of the microcapsules depends on the different techniques utilized to produce the capsules. The size distribution of the microcapsules is analyzed, for instance, with a particle size analyzer (PSA). Carrying out this method, the microcapsules are dispersed in a suitable substrate, and the particle size distribution is determined by a laser particle size analyzer [3]. Table 3-1 shows the variation in the capsule sizes in dependence on the manufacturing techniques [4].

Micro-encapsulation technique	Approximate microcapsule size range, micrometer
Spray drying	0.1-5,000
Emulsion polymerization	0.05-5
Interfacial polymerization	0.5-1,000
Coacervation	2-1,200
Sol-gel encapsulation	0.2-20

Table 3-1: Microcapsule sizes obtained by the various micro-encapsulation techniques [4, p.539].

As indicated in Table 3-1, microcapsules created by utilizing chemical methods can be made much smaller than those created in physical procedures. However, the typical sizes of microcapsules are from one micrometer to one thousand micrometers [1]. In the case of microencapsulated phase change materials, the capsule sizes are usually less than fifty micrometers.

The shell's thickness ranges from 0.1 micrometers to 0.3 micrometers. Fig. 3-3 shows microcapsules containing phase change material.

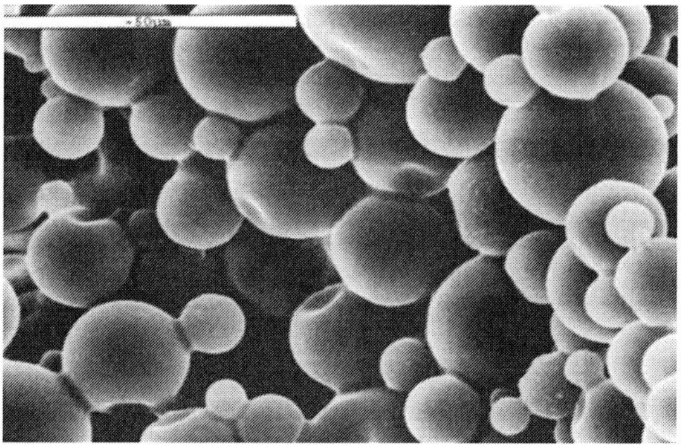

Fig. 3-3: Microcapsules containing phase change material.

Thermal stability

The thermal stability of microcapsules containing phase change material is measured with the method of thermogravimetric analysis (TGA). In the application of this measuring method, the temperature of a sample of microcapsules is increased in small intervals, and the weight loss of the test sample within each interval is determined. When paraffin is used as the core material and polymethylmethacrylate (PMMA) as the shell, the onset temperature for the decay of the paraffin is about 160 °C. The polymethylmethacrylate (PMMA) shell starts to decay at about 340 °C. The decomposition of the microcapsules is complete at about 450 °C [3].

Thermal reliability

The thermal reliability of microcapsules containing phase change material is tested in repeated thermocycles. The microcapsules are cooled below the crystallization temperature of the phase change material and then heated above the melting temperature. This procedure is carried out at least one hundred times. After the thermocycling procedure is complete, the temperature ranges of the latent heat absorption and the latent heat release, and the latent heat storage capacity of the microencapsulated phase change material, are measured with a differential scanning calorimetry (DSC) testing device. The results received are compared to the measurements obtained before the thermocycling was performed. When a sample of phase change material microcapsules was exposed to one hundred thermocycles in a trial, only a deviation of 0.5 °C in the peak melting temperature and a deviation of 1.5°C in the peak crystallization temperature were observed.

The latent heat storage capacity of the sample decreased by 2 percent in the thermocycling procedure [3, 7].

Mechanical stability

Microcapsules made for pharmaceutical applications or used in food production are produced to release their content under a specific trigger mechanism, such as mechanical pressure. In contrast, microcapsules containing phase change material should possess durable shells that prevent the core phase change material from leaking out also under mechanical forces.

Two methods are utilized to test the mechanical stability of microcapsules containing phase change material. In the first method, the microcapsules are placed between two plates. Pressure is then applied to the upper plate with the intent to break the microcapsules. On average, the microcapsules were able to withstand a pressure of about 10 kPa.

A second method consists of a shear test. The shear test is carried out in a centrifuge. The breakage rate of the microcapsules is determined from the initial mass of the microcapsules before placing them into the centrifuge and the mass of the microcapsules after the spin and is given in a percentage. The breakage rate depends on the speed at which the centrifuge operates and the duration of the spin. The breakage rate measured at a specific type of microcapsule exposed to 8,000 rounds per minute for ten minutes was determined to be about 1.6 percent [3].

Commercially available phase change material microcapsules

Commercially available microcapsules contain paraffins that operate at distinct temperatures in the temperature range between -10°C and 50°C. The microcapsule sizes vary between 5 micrometers and 30 micrometers. Microcapsules are supplied as a dry powder with a solid content above 97 percent, a cake with a solid content of about 70 percent, or a slurry with a solid content between 30 percent and 45 percent. The shell materials are mostly made of acrylic resin. The shell comprises about 20 percent of the total weight of the microcapsule. In other words, the paraffin content in the microcapsule is about 80 percent of the microcapsule's total weight.

Microencapsulation reduces the size of the paraffin droplet, decreasing the volume-based latent heat storage capacity. For instance, pure octadecane possesses a latent heat storage capacity of 244 J/g. When the octadecane is microencapsulated, the latent heat storage capacity decreases to about

180J/g. Therefore, the microcapsule shell reduces octadecane's volume-based latent heat storage capacity by about 26 percent, i.e., the shell covers about 26 percent of the microcapsule's volume.

3.4 Nano-encapsulation

Nanocapsules are structurally more stable than microcapsules. The deformation of a ten-nanometer size capsule is substantially smaller compared to the deformation of a ten-micrometer size capsule when the same force is applied. So far, phase change material nanocapsules have been solely produced under lab-scale conditions utilizing emulsion polymerization or interfacial polymerization techniques. The smallest nanocapsules obtained in the lab studies possessed sizes between 60 nanometers and 90 nanometers in diameter [2, 4, 7].

Phase change material nanocapsules seem to be specifically advantageous for applications in heat transfer fluids. When applied to these fluids, the phase change material microcapsules enhance their viscosity. As a result of the increased viscosity, the microcapsule shells can be damaged during the pumping of the fluid. The use of these smaller-sized nanocapsules would solve the problem [1].

3.5 Shape-stabilization-encapsulation

Shape-stabilization-encapsulation is another form of containment for phase change materials. Shape-stabilized-encapsulation systems durably contain the phase change material and retain their shapes even when the phase change material is in its liquid state. A shape-stabilized-encapsulation system is, for instance, made by mixing a liquid polymer such as high-density polyethylene (HDPE) with a liquid phase change material such as paraffin. Both materials are heated beyond the melting point of the high-density polyethylene (HDPE) at about 130 °C, and the mixture is thoroughly stirred. By cooling down the mixture, a polymeric phase change material hybrid is created. In its applications, the polymeric phase change material hybrid is used at temperatures above the melting temperature of the phase change material, but below the melting temperature of the polymer. Therefore, the polymer remains solid when the phase change material melts [1, 8, 9].

Another possibility for creating a shape-stabilized encapsulation system is the absorption of paraffin into a microporous granulate consisting of polypropylene powders or pellets. The polypropylene powders or pellets possess an externally accessible porosity between 70 percent and 85 percent.

In the first step of the manufacturing process, the paraffin must be melted until it is completely liquefied. Then, the micro-porous granulate is mixed into the liquefied paraffin. The paraffin is soaked into the micron-sized voids of the polymer by capillary absorption with the support of a vacuum. The strong adhesion forces, due to the large inner surface of the micro-porous cell structure, create a durable containment of the paraffin inside the granulate structure. The porous structure of the polypropylene powder or pellets allows for a loading capacity of up to 85 percent. The particle sizes range from 0.1 millimeter to 1 millimeter. A sample of the microporous polypropylene powder with an 85 percent octadecane paraffin content is shown in Fig. 3-4.

Besides the 15 percent decrease in the volume-based latent heat storage capacity of the octadecane by the enclosure, the temperature ranges of the latent heat absorption and latent heat emission of the enclosed octadecane paraffin are slightly extended, and the temperature range of the latent heat emission is shifted to slightly lower temperatures.

Fig. 3-4: Microporous polypropylene powder with octadecane paraffin contained therein.

The specific heat capacity of polypropylene is 1,700 J/kg K, which is lower than the specific heat capacity of paraffin, which ranges from 2.1 kJ/kg K to 2.9 kJ/kg K. Therefore, if paraffin is contained in a polypropylene granulate, the overall specific heat capacity drops leading to a decrease in the absorption of sensible heat [1].

The thermal conductivity of polypropylene is like the thermal conductivity of paraffin. Thus, there is no improvement in the heat transfer to and from the paraffin when contained in a polypropylene granulate.

The polypropylene powder with the octadecane paraffin contained therein has been tested in a thermocycling procedure. After one hundred thermocycles, the thermal performance was determined using the differential scanning calorimetry (DSC) testing procedure. The results were compared to those received before the thermocycling procedure was carried out. In respective tests, a thermal performance reduction of less than 3 percent was obtained.

3.6 Applications

Macroscale containers filled with phase change material are mainly utilized in cold storage and shipping systems. Microencapsulated phase change materials are applied to textiles, bedding components, medical devices, and building products. Nano-encapsulated phase change materials are especially suitable for application in heat transfer fluids. Thermal storage systems use, for instance, shape-stabilized encapsulation of phase change materials.

References

1. Fleischer, Amy, S. 2015. "Thermal energy storage using phase change materials: fundamentals and applications". Springer Briefs in Applied Science and Technology, Thermal Engineering and Applied Science, 2015 Edition, Springer Cham, Heidelberg, New York, Dordrecht, London
2. Salunkhe, Pramod B., Shembekar, Prashant S. 2012. "A review on effect of phase change material encapsulation on the thermal performance of a system". Renewable and Sustainable Energy Reviews, No. 16: 5603-5616. http://dx.doi.org/10.1016/j.rser.2012.05.037
3. Chen, Zhitong, Dou, Guijing, Hu, Yahao, Peng, Guangjian, et al. 2020. "Phase Change Material (PCM) Microcapsules for Thermal Energy Storage". Wiley Hindawi. Advances in Polymer Technology, 20 pages. https://doi.org/101155/2020/9490873

4. Farid, M., Jamekhorshid, A., Sadrameli, S.M. 2014. "A review of microencapsulation methods of phase change materials (PCMs) as a thermal energy storage (TES) medium". Renewable and Sustainable Energy Reviews, No. 31: 531-542, http://dx.doi.org/10.1016/j.rser.2013.12.033
5. Choudhury, Nitamani, Das, Kalyan, Meghwal, Murlidhar. 2021. "Microencapsulation: An overview on concepts, methods, properties, and applications in foods". Food Frontiers: 1-17, DOI: 10:1002/ fft294
6. Das, Sanjoy Kumar, David, Sheba Rani Nakka, Halder, Tripti, Khanam, Jasmina, et al. 2011. "Microencapsulation techniques and its practice". Int. Journal Pharma Science Tech, Vol. 6, No. 2: 1-23
7. Alkan, Cemil, Bicer, Alper, Dögüscü, Derja Kahraman, Sari, Ahmet. 2014. "Micro/nano-encapsulated *n*-heptadecane with polystyrene shell for latent heat thermal energy storage". Solar Energy Materials & Solar Cells, No. 126: 42-50, http://dx.doi.org/10.1016/j.solmat.2014.03.023
8. Linganiso, Linda Z, Mochane, Mokgaotsa J., Mokhena, Teboho C., Motaung, Tshwafo E. 2020. "Shape-stabilized phase change materials of polyolefin/wax blends and their composites". Journal of Thermal Analysis and Calorimetry, No. 139: 2951-2963 https://doi.org/10.1007/s10973-019-08734-3
9. Salyer, Ival O. 1990. "Method for preparing polyolefin composites containing a phase change material". US Patent No. 4,908,166

CHAPTER FOUR

CONTAINMENT

4.1. Forms of containment

Containment is different from encapsulation. Encapsulation durably encloses the phase change material, while the containment only surrounds the phase change materials. Under certain circumstances, some of the micro-encapsulated/nano-encapsulated or non-encapsulated phase change materials will diffuse out of the containment matrix.

Micro/nano-encapsulated phase change materials, as well as non-encapsulated phase change materials, are, for instance, contained in textiles, polymers, or combinations of them for use in various applications. Such containments comprise:

- Fibers and yarns made from them.
- Nonwovens.
- Foams.
- Coatings.
- Polymeric films.

4.2. Fibers and yarns

Acrylic fibers

The containment of microencapsulated phase change materials in fibers started with acrylic fibers. Therefore, this fiber type should be used to explain the technology in more detail. The concept of utilizing microencapsulated phase change materials inside fibers and, therefore, creating fibers with reversible enhanced thermal storage properties was invented by Bryant and Colvin from Triangle Research Institute in Rayleigh, North Carolina, USA in the 1980s [1].

In 1988, microcapsules containing eicosane-phase change material were first integrated into acrylic fibers using a wet spinning technique [2]. In the wet spinning process utilized to create acrylic fibers, the polymer polyacrylonitrile is first dissolved in a spinning solvent such as aqueous sodium thiocyanate. This solution is called dope. The polymer dope is then extruded through a spinneret that has a significant number of capillaries. The diameters of the capillaries range from 0.05 millimeters to about 0.25 millimeters, depending on the desired diameters of the fibers. The polymer solution is pumped through the spinneret into a liquid bath containing a coagulant which is a nonsolvent. In the coagulation bath, the polymer is precipitated by the counter diffusion between the solvent in the polymer solution and the nonsolvent. In this way, a solid gel filament is formed which is taken out of the coagulation bath by rolling over the first godet. The jet stretch, i.e., the ratio of roll surface speed to theoretical extrusion velocity, creates tension leading to the first orientation of the polymer chain. Further orientation takes place when the coagulated filaments are washed free of solvent and stretched in steam or hot water. After washing and stretching, a finish is applied to provide surface lubrication for further fiber processing and to prevent static buildup. The fiber bundle is then dried and collected.

In manufacturing acrylic fibers with incorporated microencapsulated phase change material, the microcapsules are already added to the polymer solution in the first step of the spinning process. The portion of microcapsules in the polymer spinning dope is, for instance, about 30 percent on a weight basis. Therefore, the microcapsules containing the phase change material undergo all the steps of the spinning process and must resist mechanical interaction as well as temperature changes. At the end of the process, the phase change material microcapsules are surrounded by the fiber material and permanently locked inside the fibers [3]. Fig. 4-1 shows a cross-section through acrylic fibers with incorporated phase change material microcapsules.

To produce acrylic fibers with phase change material microcapsules of sufficient quality, some requirements must be considered:

- The microcapsule size needs to be less than 10 micrometers.
- The microcapsules must be thoroughly mixed into the main polymer stream before the extrusion process is carried out.
- The mixture of the acrylic polymer with the microcapsules needs to be filtered before the fiber spinning process to prevent the holes in the spinnerets from blocking.

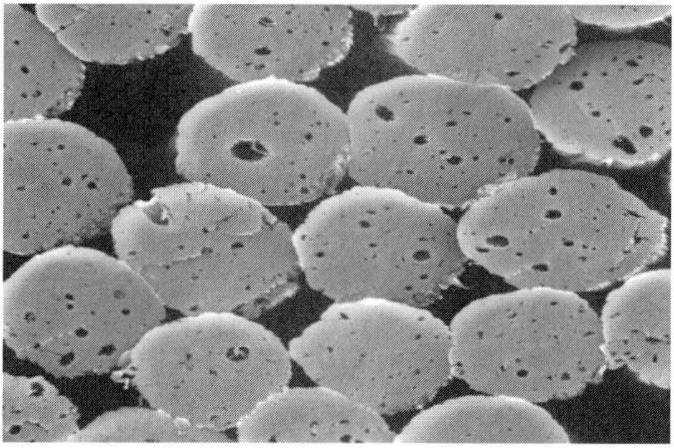

Fig. 4-1: Cross-section through OUTLAST® acrylic fibers (source: Outlast Technologies, Inc.).

Acrylic fibers generally contain a microcapsule quantity between 5 wt.% and 10 wt.%. A maximum loading level of 15 wt.% is possible without losing a significant proportion of the fiber strength. A microcapsule quantity between 5 wt.% and 10 wt.% provides a latent heat storage capacity between 3 J/g and 7 J/g. The sensible heat storage capacity of an acrylic fiber covers the range of 1.2 J/g K to 1.5 J/g K. Therefore, by adding phase change material microcapsules to the acrylic fiber matrix, their heat storage capacity can be enhanced by factors between 2 and 5 in the temperature range where the phase change takes place. The latent heat storage capacity increases proportionally to the increase in the phase change material microcapsule content in the fibers, as shown in Fig. 4-2. The results shown in Fig. 4-2 are based on the same fiber size of 5.5 dtex.

A fiber's linear density is given in denier (den) or dtex. This measurement represents the weight of the fiber in reference to the length of the fiber.

$$1 \text{ den} = 1\text{g} / 9000 \text{ m} = 0.11 \text{ mg/m} \tag{4.1}$$

$$1 \text{ dtex} = 1 \text{ g} / 10{,}000 \text{ m} \tag{4.2}$$

$$1 \text{ dtex} = 1 \text{ den} / 0.9 \tag{4.3}$$

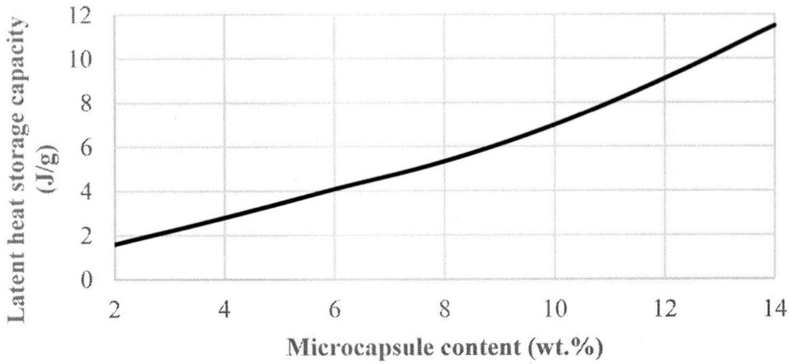

Fig. 4-2: Latent heat storage capacity in dependence on the acrylic fiber's microcapsule content.

The acrylic fibers with incorporated phase change material microcapsules are manufactured in different sizes. With an increase in the fiber size, the microcapsule content in the fiber rises, and therefore, the latent heat storage capacity of the fiber increases. Fig. 4-3 shows the latent heat storage capacity increase in dependence on the fiber's linear densities.

The latent heat storage capacity rise is proportional to the increase in the fiber's linear density. For instance, an increase in the fiber's linear density from 2.2 dtex to 3.3 dtex enhances the latent heat storage capacity by about 1 J/g. The fibers utilized in this test were equipped with a phase change material microcapsule content of about 9 wt.%.

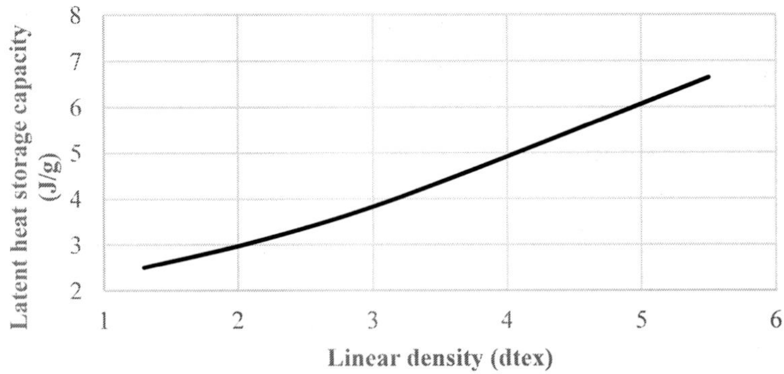

Fig. 4-3: Latent heat storage capacity in dependence on the fiber's linear density.

With an increase in the fiber's linear density, the melting temperature range of the incorporated phase change material noticeably changes, which needs to be considered in the fiber's applications. Fig. 4-4 illustrates that the melting temperature range of the incorporated phase change material extends slightly and shifts to lower temperatures with an increase in the fiber's linear density.

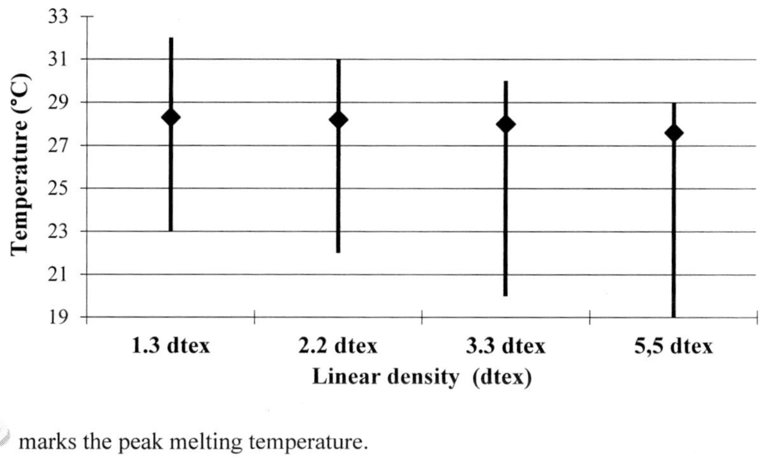

Fig. 4-4: Change in the melting temperature range with a change in the fiber's linear density.

The acrylic fibers with incorporated microencapsulated phase change material have been produced and commercialized as OUTLAST® acrylic fibers. These fibers have been made in different linear densities. OUTLAST® acrylic fibers are staple fibers made in stable lengths between 28 millimeters and 120 millimeters. Several phase change material microcapsules operating in different temperature ranges have been utilized in manufacturing OUTLAST® acrylic fibers. The phase change material microcapsules have been applied to the fibers in quantities between 5 wt.% and 9 wt.%.

OUTLAST® acrylic fibers were tested compared to regular acrylic fibers at the Henan Institute of Engineering in Zhengzhou, China by Haixia Zhang and Xichang Zhang [4]. These researchers tested OUTLAST® acrylic fibers with a stable length of 38 millimeters and a linear density of 1.9 dtex. The test fibers possessed a microcapsule content of 7 wt.%. Differential scanning calorimetry (DSC) testing results indicated that the latent heat absorption of the phase change material utilized in the fibers took place in a temperature range between about 20 °C and 30 °C with a single peak at a temperature of 26 °C. The latent heat release covered a temperature range from 10 °C to 30 °C with a single peak at a temperature of 22 °C. The latent heat storage capacity was determined to be about 5 J/g.

The Chinese researchers measured the decomposition temperature of the OUTLAST® acrylic fibers using thermogravimetric analysis (TGA) and determined it to be about 311 °C.

The reference material utilized in the comparison was a regular acrylic fiber with a stable length of 36 millimeters and a linear density of 1.5 dtex. The researchers from the Henan Institute of Engineering in Zhengzhou, China, measured the tensile strength, elongation, number of crimps, elastic recovery of crimps, and moisture regain, among other properties. The test results are summarized in Table 4-1.

Properties	OUTLAST® acrylic fiber	Regular acrylic fiber
Tensile strength, cN/dtex	2.46	2.98
Elongation, %	29.95	22.51
Number of crimps, crimps /cm	3.16	2
Elastic recovery of crimp, %	76.07	84.12
Moisture regain, %	3	1.6

Table 4-1: Physical properties of OUTLAST® acrylic fibers and regular acrylic fibers determined at the Henan Institute of Engineering in Zhengzhou, China.

The test results summarized in Table 4-1 indicate that in comparison to the regular acrylic fibers, the OUTLAST® acrylic fibers possess a lower tensile strength and a higher elongation. OUTLAST® acrylic fibers exceeded the regular acrylic fibers in the number of crimps and in the moisture regain. The elastic recovery of crimps is lower for the OUTLAST® acrylic fibers than that of regular acrylic fibers.

Yarns made of OUTLAST® acrylic fibers

OUTLAST® acrylic yarns are made from 100 percent OUTLAST® acrylic fibers. There are no certain requirements regarding the yarn spinning process utilized in making OUTLAST® acrylic yarns. For instance, ring, open-end, or air jet spinning systems can be applied, as well as cotton, worsted, or woolen systems. One spinning system used to produce OUTLAST® acrylic yarns is the worsted system.

Using the worsted system, longer stable fibers over 75 millimeters are usually processed. They possess linear densities between 2.2 dtex and 3.3dtex.

In the worsted yarn spinning system, loose fibers of different bales are first blended and formed into a picker lap which goes into the carding machine. On the carding machine, fine bent wires on revolving cylinders pull the fibers apart, remove waste, and arrange the fibers so they can be spun into yarn. The fibers leave the carding machine as a fine web gathered into a loose rope called a sliver. The sliver is then coiled in cans.

After that, shorter fibers are removed from the sliver in a combing procedure, and the remaining long fibers are arranged parallel to each other.

In the next step, the sliver runs between sets of rollers, each moving faster than the one before. This results in several slivers drawn out to the thickness of one sliver. The procedure is repeated until the fibers are well mixed into a roving.

In the consecutive spinning process, the twist is inserted as the fibers from the roving are carried by the traveler around the edge of the ring, inside of which is the faster rotating spindle. The sliver is drawn out to yarn size in this production step. It is finally wound up on the fast-moving spindle called "kopse" (Fig. 4-5). After the spinning process, the yarn is steamed to take out some of the yarn's twist.

Fig. 4-5: Yarn spinning machinery (source: Schoeller GmbH & CoKG, Austria).

In the final production step, the yarn is winded off from the 'kopse' with a weight of 125 grams to a spool with a weight of 750 grams. Then two or more yarns are twisted together to form a twisted yarn.

Tests of the yarns

Measurements of the latent heat storage capacities of samples taken at different stages of the production process have indicated that up to 50 percent of the latent heat storage capacity was lost, probably due to a loss in phase change material microcapsules during the spinning process. This loss must be considered when knits or woven fabrics are made from the yarn.

Comparison tests with regular acrylic yarns indicated similar results regarding elongations and a slightly lower straight tenacity.

Knits made of OUTLAST® acrylic yarn

Knits are made from 100 percent OUTLAST® acrylic yarns or by blending the OUTLAST® acrylic yarns with a yarn of another fiber such as wool. It needs to be considered that making a knit by blending the OUTLAST® acrylic yarn with another yarn always leads to a substantial loss in latent heat storage capacity and, hence, in thermal performance. Therefore, from a thermal performance point of view, it is more sufficient to make the knits out of 100 percent OUTLAST® acrylic yarn. A knit with a weight of 250g/m² and a thickness of one millimeter absorbs, stores, and releases a latent heat of about 1 kJ/m². Doubling weight and thickness of the knit also doubles its latent heat storage capacity.

Knits support the latent heat absorption and the latent heat release of the phase change material. There is also no loss in the water vapor permeability and hence, breathability than normal acrylic yarns are substituted by the OUTLAST® acrylic yarn in making knits.

Tests of different versions of the OUTLAST® acrylic yarn have indicated that there is also a correlation between latent heat storage capacity and yarn's linear density, as shown in Fig. 4-6.

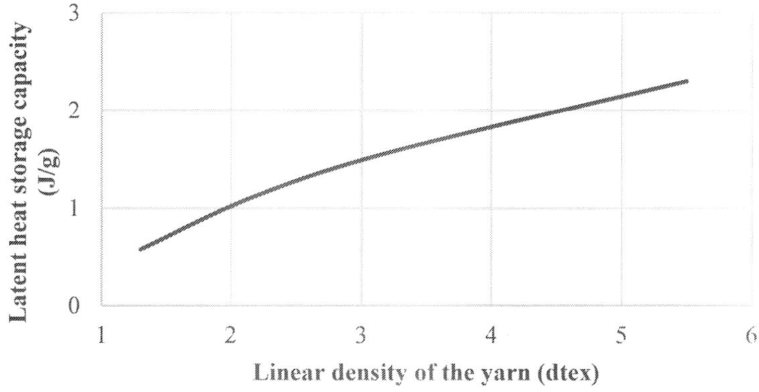

Fig. 4-6: Latent heat storage capacity in dependence on the yarn's linear density.

As indicated in Fig. 4-6, the latent heat storage capacity of knits can not only be enhanced by making the knits heavier and thicker but also by using yarns of larger sizes. On the other side, no significant differences have been experienced in latent heat storage capacities received in tests of different knitting types based on similar fabric weights.

Viscose rayon fibers

OUTLAST Technologies, Inc. announced in 2006 the manufacturing of cellulosic viscose fibers containing microencapsulated phase change material.

Viscose rayon fibers with incorporated microencapsulated phase change material are manufactured in a solution-spinning process [5]. The basic cellulosic material can consist of sheets of cellulose, wood pulp, or cotton linters. The cellulosic material is first immersed into a basic solution material, squeezed through rollers, and then shredded to form crumbs. The crumbs are then treated with carbon disulfate to create cellulose xanthate.

In the next steps, the crumbs are dissolved in a solvent to form a viscous solution. The microencapsulated phase change material is added to the viscous solution and mixed therein before passing through the orifices of the spinneret. After the transfer through the spinneret, the viscose rayon fibers are washed, dried, drawn, and stretched. A cross-section of the Outlast® viscose rayon fiber is shown in Fig. 4-7.

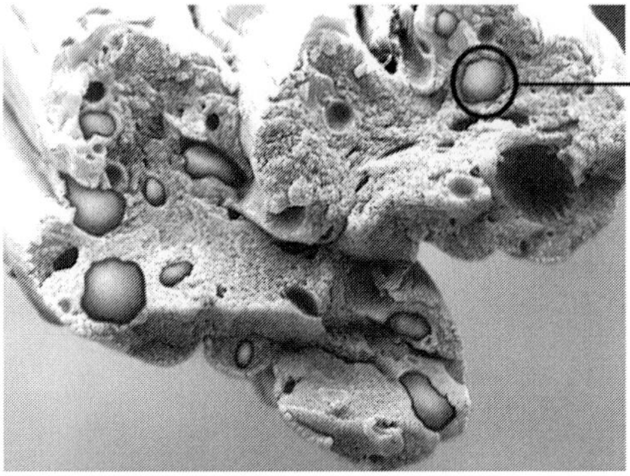

Fig. 4-7: Cross-section through the Outlast® viscose rayon fiber (source: Outlast Technologies, Inc.).

Outlast® viscose rayon stable fibers, for instance, have a nominal stable length of 70 millimeters, and a linear density of 7.0 dtex. For these fibers, a minimum tensile strength of 6 cN/tex and an elongation of 15 percent was measured.

Tests of Outlast® viscose rayon fibers

Outlast® viscose rayon fibers were tested at the Henan Institute of Engineering in Zhengzhou, China, by Ruichao Xu, Haixia Zhang, and Xichang Zhang, [6]. These researchers tested Outlast® viscose rayon fibers with a stable length of 40 millimeters and a linear density of 1.7 dtex. The test fibers possessed a microcapsule content of 7 wt.%.

The researchers measured the thermal performance of the Outlast® viscose rayon fibers with a differential scanning calorimetry (DSC) testing device. The test results received in the heating process showed a latent heat absorption in a temperature range between 15 °C and 35 °C with a double peak at about 26 °C and 30 °C. In the cooling process, a double peak regarding the latent heat release was observed at about 13 °C and 18°C. The entire latent heat release covered a temperature range between 10 °C and 30°C. The latent heat storage capacity was measured to be about 11 J/g.

The decomposition temperature was measured with a thermogravimetric analysis (TGA) testing device and determined to be between 280 °C and 320 °C.

Polyester fibers

Polyester fibers, as well as nylon fibers or polypropylene fibers, are manufactured in a melt-spinning process [7]. The melt spinning is carried out in a temperature range between 200 °C and 380 °C. The process pressure ranges up to 3000 psi. The high temperatures, the high pressures, and the duration over which the phase change material microcapsules are exposed to these temperatures and pressures in the melt spinning process would lead to a decomposition of the microcapsules [8]. Therefore, another solution for producing polyester fibers with microencapsulated phase change materials had to be established.

One solution proposed by Hartmann was the improvement of the microcapsule-phase change material system [9]. A two-cavity system was created where the phase change material is integrated into an internal cavity and, a stabilizing agent is arranged in the external cavity.

A second solution, developed by Brang et al., consists of a bi-component fiber, where phase change material is contained in an active core and an outer sheath comprises the fiber material such as polyester [10]. In this approach, microencapsulated phase change material in the form of a wet cake is mixed into the molten polymer creating the core of the fibers. The resulting first blend is cooled and chopped for further processing. This first blend is then mixed with a second polymer which later creates the outer sheath of the fibers. The mixture is put through an extruder to form pellets which need to be dried before the spinning process. In the melt-spinning process, the pellets are first melted to form a syrup-like solution. The solution is put in a metal container, the spinneret, and forced through orifices to form the bi-component fibers.

Outlast ® polyester stable fibers are manufactured in variants with a nominal stable length of 24 millimeters and a linear density of 2.2 dtex, as well as with a nominal stable length of 60 millimeters and a linear density of 6.7dtex. The technical properties of these stable fibers are summarized in Table 4-2.

Outlast® polyester stable fibers absorb latent between 26 °C and 33 °C. They possess a latent heat storage capacity of about 8 J/g [11, 12].

Properties	2.2 dtex OUTLAST® polyester stable fiber	6.7 dtex OUTLAST® polyester stable fiber
Tensile strength, cN/dtex	31	24
Elongation, %	50	20

Table 4-2: Technical properties of OUTLAST® polyester stable fibers.

Polypropylene fibers

Polypropylene fibers incorporating phase change material were developed in the lab in recent years. For instance, hollow propylene fibers were filled by Hufenus et al. with polyethylene glycol phase change material utilizing a microfluidic filling technology [13]. The phase change material filling reached a maximum content of 84 wt.%. However, the formation of bubbles inside the fibers could not be prevented reducing the thermal efficiency of the fiber material. The latent heat storage capacity remained stable in repeated thermocycles.

Iqbal and Sun developed polypropylene fibers with microencapsulated phase change material at Heriot-Watt University in the United Kingdom [14, 15]. Microencapsulated octadecane was mixed with polypropylene granulate in a container and well shaken to obtain a homogenous mixture. Mineral oil was added during the mixing process to obtain an even dispersion of the microcapsules in the polypropylene granulate. The melt-spinning process was carried out with an extruder, and a winder collected the filament through a water bath. To reduce the duration in the extrusion chamber and to avoid damaging the microcapsules due to high temperatures and pressure, the spinneret used in the process was modified decreasing the number of holes and increasing their diameter. After the spinning process, the filaments were drawn out on a drawing machine to enhance their strength and obtain the desired fineness of the fiber. A maximum microcapsule concentration of 12 wt. % was obtained leading to a latent storage capacity of 9.2 J/g. The strength of the fibers was also tested, and a tenacity of 37 cN/tex was obtained.

4.3. Nonwovens

Nonwovens are made from a web of fibers. For instance, a polyester nonwoven is created from a blend of regular polyester fibers with a linear density between 2.2 dtex and 5.5 dtex and a stable length of 38 millimeters.

Manufacturing
There are two major steps in the manufacturing of nonwovens: the web formation and the bonding of the fibers in the web [16].

Web formation

The web of the polyester fibers is created in a dry-laid process. In this process, rotating rollers are used to card the fibers into a unidirectional web.

Bonding of the fibers in the web

A needle-punching procedure is used to pre-bond the web. In this procedure, barbed needles are pushed through the web. The barbed needles catch the fibers on the surface of the web and push them into the center. In this way, the web structure is densified, and the strength of the web is created through entanglement. The needling can be done on both sides of the web.

The web is then submerged in a binder bath to bond the fibers. For instance, a self-cross-linking acrylic binder is used containing phase change material microcapsules in a quantity of 45 wt.%. After the binder bath, the web is dried in a dryer at a temperature of 160 °C.

Tests

The polyester nonwovens treated with microencapsulated phase change materials were made in different weights between 50 g/m² and 200 g/m². The structure of the nonwovens also varied leading to differences in their densities. The latent heat storage capacity of the polyester nonwovens with microencapsulated phase change material rises with an increase in weight and density as shown in Fig. 4-8.

The rise in the latent heat storage capacity of the nonwoven materials is due to the increase in the microcapsule quantity per unit area.

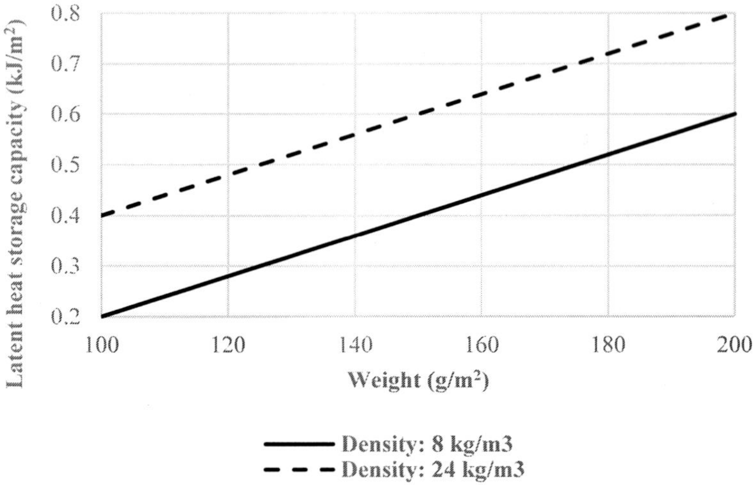

Fig. 4-8: Latent heat storage capacity of the nonwovens in dependence on weight and density.

4.4 Foams

In another application, the microencapsulated phase change material is incorporated into a flexible polyurethane foam.

Manufacturing

Before the foaming process takes place, the microencapsulated phase change material is mixed into the polyurethane solution. In the process, the microcapsules are made wet and equally dispersed throughout the mixture. Furthermore, a hardener is added to the solution causing the chemical reaction. Then, the foaming process consists of either thermally setting the polymer matrix with heat or bubbling a gas through the polyurethane solution while hardening.

The wetting process coats the microcapsules. This coating is maintained around the microcapsules during and after foaming due to the surface tension of the polyurethane matrix. The surface tension also prevents the microcapsules from crossing the boundary into the gas bubbles. That means that the gas pockets formed during the foaming process are free of microcapsules [17]. Fig. 4-9 shows a cross-section of polyurethane foam with incorporated phase change material microcapsules.

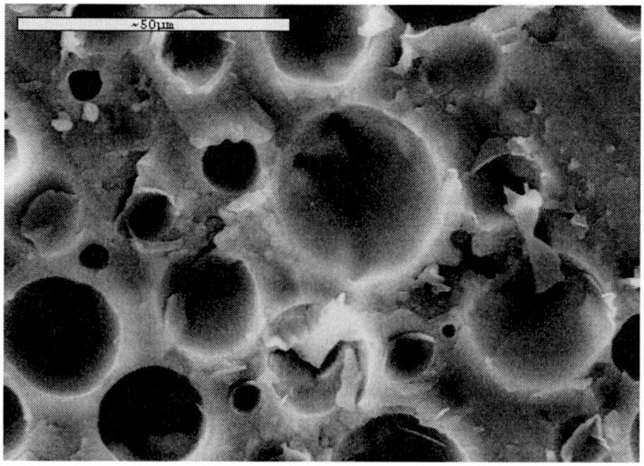

Fig. 4-9: Cross-section of polyurethane foam with microencapsulated phase change material incorporated therein.

Microcapsules with phase change material operating in different temperature ranges were applied to several polyurethane foams. Sometimes they were mixed to cover a broader temperature range.

Tests

Flexible polyurethane foam with a weight of about 570 g/m² has been loaded with microcapsules containing a phase change material with a melting point of 28 °C and a latent heat storage capacity of about 185 J/g in quantities between 100 g/m² and 250 g/m². The incorporation of the micro-encapsulated phase change material increases the foam's weight by between 15 percent and 30 percent. The rise in the latent heat storage capacity in dependence on the loading level of the microcapsules is shown in Fig. 4-10.

The increase in the microcapsule content of the polyurethane foam results in a rise in its density by about 32 percent and its thickness by about 10 percent. Due to the thickness and density increase, the foam's water vapor permeability decreases (Fig.4-11).

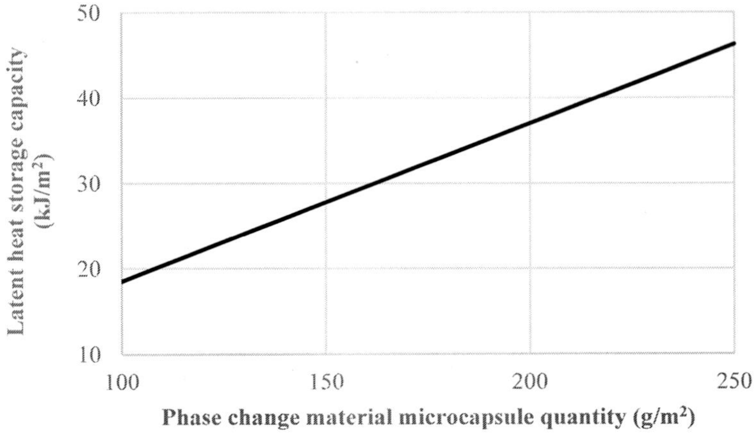

Fig. 4-10: Latent heat storage capacity in dependence on the microcapsule quantity applied to the polyurethane foam.

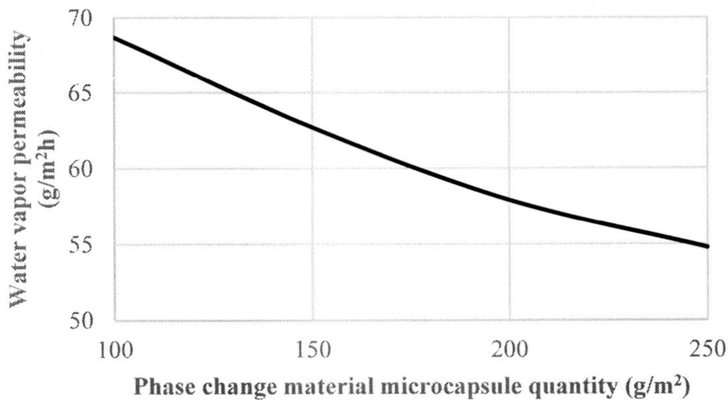

Fig. 4-11: Water vapor permeability of the foam in dependence on the microcapsule quantity.

In contrast to the reduction in its water vapor permeability, the foam absorbs an increased amount of moisture with the application of the phase change material microcapsules. For instance, by adding a microcapsule quantity of about 200 g/m^2 to the foam, its moisture absorption increased by about 25 percent.

4.5 Coatings

Microencapsulated phase change material and non-microencapsulated phase change material can also be applied to woven fabrics, knitted fabrics, nonwoven fabrics, or foams as a coating layer [18].

Manufacturing

To perform such a coating procedure, the microencapsulated phase change material or the non-microencapsulated phase change material is stirred into a binder. To avoid dissipation of the non-microencapsulated phase change material out of the binder when in its liquid stage, the phase change material needs to be cross-linked into the binder matrix. Common polymer binders used for coatings of substrates are acrylic binders, polyurethane binders, rubber latex binders, and silicone rubber binders. Typical concentrations of microencapsulated phase change material or non-microencapsulated phase change materials added to the polymer binder range from 30 wt.% to about 60 wt.%. Fig. 4-12 shows phase change material microcapsules in an acrylic coating compound applied to a fabric.

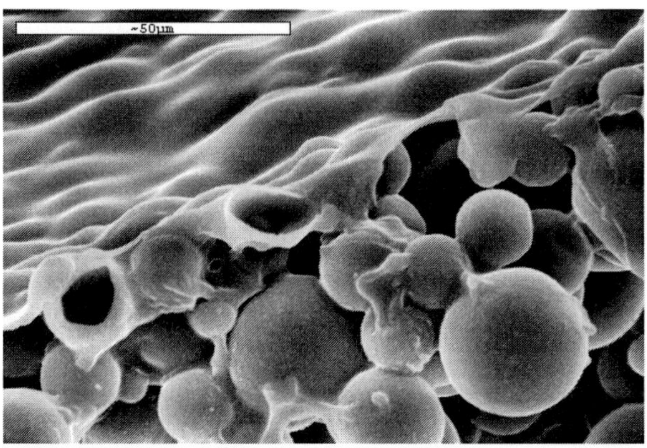

Fig. 4-12: Phase change material microcapsules incorporated into an acrylic coating compound applied to a fabric.

In applying the polymer binder with the incorporated microencapsulated phase change material or the non-microencapsulated phase change material to a fabric or foam, the knife-over-roll coating method is widely used. In the knife-over-roll coating procedure, the viscous binder is spread onto the fabric's surface or the foam's surface and is passed under the coating knife. The knife serves to spread the binder uniformly across the entire width of the substrate. In the same way, the weight of the binder applied to the substrate is controlled [19]. A knife-over-roll coating equipment is shown in Fig. 4-13.

Fig. 4-13: Knife-over-roll coating equipment.

Once coated, the substrate is passed through an oven, where the resin is cured within a few minutes. Curing temperatures typically range from 80°C to 160 °C. Besides knife-over-roll coating, other suitable coating procedures are knife-over-air coating, transfer/cast coating, dip coating, or spray coating [20].

In most applications, where a fabric is coated with a binder containing microencapsulated or non-microencapsulated phase change material, the fabric's surface is completely covered with the coating layer as shown in Fig. 4-14. When a coating layer covers the entire surface of a fabric, water vapor transfer through the fabric is reduced, decreasing the fabric's breathability [21].

To avoid this problem, screen-printing technology can be used instead. In this technology, the polymeric binder with the microencapsulated phase change material or the non-microencapsulated phase change material is applied to the fabric's surface, for instance, in discrete dots or waves. This technique allows for a substantially higher water vapor transfer through the fabric, enhancing its breathability. Compared to a coated fabric, the screen-printed fabric is softer and more flexible. However, a disadvantage of utilizing this technique is the decrease in latent heat storage capacity due to a reduction in the amount of phase change material applied. An example of a fabric that is screen-printed with a binder containing non-microencapsulated phase change material is shown in Fig. 4-15.

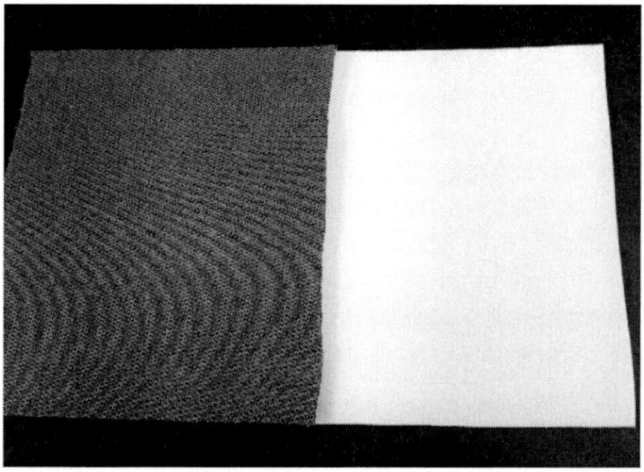

Fig. 4-14: Uncoated side of the fabric (left) and coated side of the fabric (right).

Fig. 4-15: Screen-printed fabric.

The amount of polymeric binder and, hence, the amount of microencapsulated or non-microencapsulated phase changer material applied to a textile substrate depends on its structure and thickness. The thinner the fabric, the lower the quantity of coating material that can be applied.

Coating compounds with phase change material microcapsule quantities between 30 g/m² and 100 g/m² are usually applied to thin woven fabrics. Microcapsule quantities up to about 200 g/m² are applied in coatings of foams. The thermal capacity of a thin woven fabric is enhanced substantially even when it is coated only with a small amount of microencapsulated phase change material.

Tests

The latent heat storage capacities of fabrics dependent on the applied amount of phase change material microcapsules were evaluated in tests. Fig. 4-16 shows the increase in the latent heat storage capacity of a 150 g/m² woven fabric coated with an acrylic binder containing microencapsulated phase change material in amounts between 30 g/m² and 60g/m². Adding a quantity of 60 g/m² of microencapsulated phase change material to the acrylic coating compound enhances the woven fabric's latent heat storage capacity by a factor of 7.

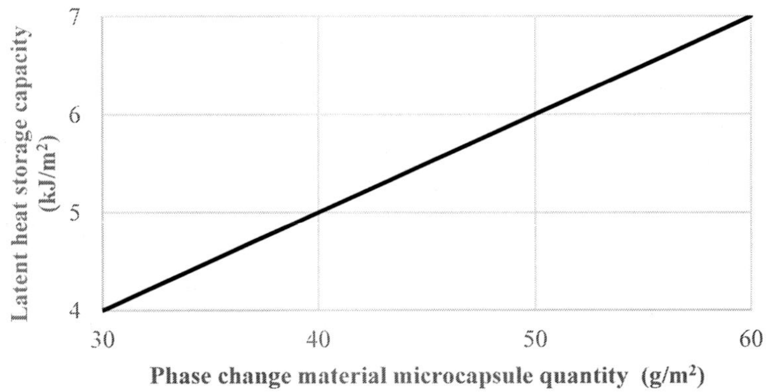

Fig. 4-16: Latent heat storage capacity in dependence on the applied quantity of microencapsulated phase change material.

By adding 60 g/m² microcapsules with phase change material embedded in the coating compound, the fabric's weight has increased by about 60 percent. In addition to that, the thickness was enhanced by about 5 percent, and the density rose by about 50 percent. Further tests of the fabric indicated an increase in the fabric's stiffness and a decrease in the fabric's water vapor permeability (breathability). For instance, applying about 100 g/m² of acrylic coating compound, which includes about 60 g/m² phase change material microcapsules to the fabric decreases its water vapor permeability by nearly half. Some test results are summarized in Fig. 4-17.

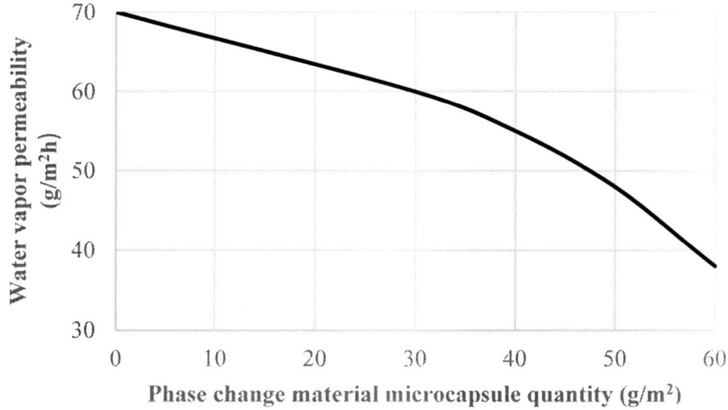

Fig.4-17: Water vapor permeability of the coated fabrics in dependence on the applied phase change material microcapsule quantity.

4.5 Polymeric films

Microencapsulated phase change material and non-microencapsulated phase change material are incorporated into polymeric films. One example of such a polymeric film is made of silicone rubber to which especially non-microencapsulated phase change materials are applied [22].

These solid-liquid phase change materials need to be cross-linked into the polymeric matrix to prevent dissolution while they are in their liquid stage. This technology creates a variety of new possibilities for the application of phase change material technology. Besides paraffin, the opportunity exists to utilize a wide range of salt hydrates, eutectics, non-paraffinic organics, and solid-solid phase change materials in such polymeric films.

Manufacturing

In creating the polymeric matrix, a two-component silicone rubber compound is preferably used. The phase change material is first mixed into the base component of the silicone rubber compound. In the second step, the catalyst is added to the mixture. The system is either extruded into a thin film or molded into patches and then cured under heat exposure. A silicone rubber film with incorporated phase change material is shown in Fig. 4-18.

Fig. 4-18: Silicone rubber film with phase change material.

The film with the incorporated phase change material can be laminated to a substrate to create a composite. Patches of the film can be quilted between two fabric layers as shown in Fig. 4-19.

Fig. 4-19: Silicone rubber film patches with phase change material quilted between two fabric layers.

References

1. Bryant, Yvonne G., Colvin, David P. 1988. "Fiber with reversible enhanced thermal storage properties and fabrics made therefrom." Patent No. US 4,756,958
2. Catarino, Andre, Onofrei, Elena, Rocha, Ana Maria. 2010. "Textiles integrating PCMs – a review". Buletinul Institutului Polittehnic DIN IASI, Tomul LVI (LX) f. .2: 99-110

3. Ashraf, Munir, Basit, Abdul, Iqbal, Kashif, Khan, Asfandyar, et al. 2019. "Phase change materials, their synthesis and application in textiles – a review". The Journal of The Textile Institute, https:// doi.org/10.1080/00405000.2018.1548088
4. Zhang, Haixia, Zhang, Xichang. 2013. "Study on Characteristics of Outlast Acrylic Fiber". Advanced Material Research, Vol. 627: 85-89,
5. Hartmann, Mark, North, Matthew, Worley, James Brice. 2019. "Cellulosic fibers having enhanced reversible thermal properties and methods of forming thereof ". Patent No. US 10,208,403
6. Xu, Ruichao, Zhang, Haixia, Zhang, Xichang. 2011. "Structure and Thermo-Regulated Performance of Outlast Viscose Fiber". Advanced Material Research, Vol. 332-334: 812-815
7. Mondal, Subrata. 2020. Phase Change Fibers. Handbook of Fibrous Materials, First Edition, Wiley-VCH Verlag GmbH: 263-279
8. Lipol, Lafayet Sulfari. 2022. "Characterization of PCM fibers". Textile Today, November 25, 2022
9. Hartmann, Mark H. 2004. "Stable phase change materials for use in temperature regulating synthetic fibers, fabrics, and textiles". Patent No. US 6,689,466
10. Brang, James E., Haggard, Jeffrey S., Hartmann, Mark, Magill, Monte Christopher. 2014 "Multi-component fibers having enhanced reversible thermal properties and methods of manufacturing thereof". Patent No. US 8,679,627
11. Outlast Technologies GmbH. 2021. Product information about Outlast ® 2.2 dtex stable polyester fiber
12. Outlast Technologies GmbH. 2021. Product information about Outlast ® 6.7 dtex stable polyester fiber.
13. Hufenus, Rudolf, Li, Weipei, Lin, Chao, Yan, Yurong, et al. 2021. "Flexible Phase Change Material Fiber: A Simple Route to Thermal Energy Control Textiles". Materials No. 14, 401, https:// doi.org/10.3390/ma1420401
14. Iqbal, K., Sun, D. 2014. "Development of thermo-regulating polypropylene fibre containing microencapsulated phase change materials". Renewable Energy, Vol. 71: 473-479, DOI: 10.1016/j.renene.2014.05.063
15. Iqbal, K., Sun, D. 2015. "Development of thermal stable multifilament yarn containing micro-encapsulated phase change materials". Fibers and Polymers, Vol. 16. No.5: 1156-1162, DOI: 10.1007/s12221-015-1156-9

16. Dietel, David S., Grynaeus, Peter, Johnson, Susan Gwynneth, O'Regan, Terry, et al. 2013. "Thermal control nonwoven material". Patent No. US 8,449,947
17. Beretta, Elisa M., De Campos, Tatiana L. Avila, Kindlein, Wilson, et al. 2016. "Engineering design: Eicosane microcapsules synthesis and application in polyurethane foams aiming to diminish wheelchair cushion effect on skin temperature". Journal of Engineering Science and Technology, Vol. 11. No. 12: 1818-1834
18. Mondal, S. 2008. "Phase change materials for smart textiles – An overview". Elsevier, Applied Thermal Engineering, Vol. 28: 1536-1550, DOI: 10.1016/j.applthermaleng.2007.08.009
19. Pause, Barbara. 2017. "Smart Textiles Created by the Application of Phase Change Material (PCM)". ITMC International Conference on Intelligent Textiles and Mass Customization, Ghent, Belgium, October 15-17, 2017
20. Pause, Barbara. 2010. "Phase change materials and their application in coatings and laminates for textiles". In: Smart textile coatings and laminates, edited by Smith, W.C., 236-250. Woodhead Publishing Ltd.
21. Shin, Younsook, Son, Kyunghee, Yoo, Dong-Il. 2005. "Development of Thermoregulating Textile Materials with Microencapsulated Phase Change Materials (PCM). IV. Performance Properties and Hand of Fabrics Treated with PCM Microcapsules". Journal of Applied Polymeric Science, Vol. 97: 910-915, DOI: 10.1002/app.21846
22. Pause, Barbara Hildegard. 2009. "Material made from a silicone rubber, production process, and application". Patent No. US7,488,773

CHAPTER FIVE

THERMAL EFFECTS

5.1. Overview

Nowadays, a significant portion of phase change material technology is used to store thermal energy. The thermal effect consists of absorbing latent heat, storing it, and utilizing the stored latent heat when needed.

However, several other thermal effects are created by applying phase change material, which a product or a system can benefit from. These effects comprise:

- Enhancement of the heat storage capacity.
- Cooling effect.
- Heating effect.
- Thermoregulating effect.
- Active thermal barrier effect.

All these effects improve the thermal performance of the prevailing product or system. However, some principles must be considered and implemented to utilize these thermal effects properly [1, 2].

5.2. Thermal energy storage

Many large-scale solar power plants produce a surplus of energy during the day. If some of the excess energy can be diverted into a storage system, stored as thermal energy, and utilized overnight or the next day, the plant can operate more efficiently.

The storage system usually consists of large storage tanks filled with solid phase change material, which transitions into its liquid state when thermal energy is absorbed. The thermal energy is stored in the liquid phase change material and can be extracted by solidifying the phase change material again. However, an obstacle to the efficient operation of such a system is the low thermal conductivity of a variety of phase change materials, which will slow down the thermal energy exchange substantially [3, 4]. Methods to enhance the heat transfer to and from the phase change material will be introduced and discussed in Chapter Seven.

5.3. Enhancement of the heat storage capacity

Every material absorbs sensible heat in a heating process, leading to a constant rise in temperature. For instance, a one-kilogram silicone patch absorbs about 0.7 kJ in sensible heat while its temperature rises by one degree.

Incorporating the salt hydrate lithium nitrate trihydrate in a quantity of 40wt.% into the silicone patch leads to a latent heat storage capacity of about 120 kJ. To absorb the same amount of heat as the original silicone patch without the phase change material, the patch would need to be heated until its temperature is raised by about 170 degrees Celsius. Fig. 5-1 compares the heat storage capacity of the silicone patch with and without the phase change material.

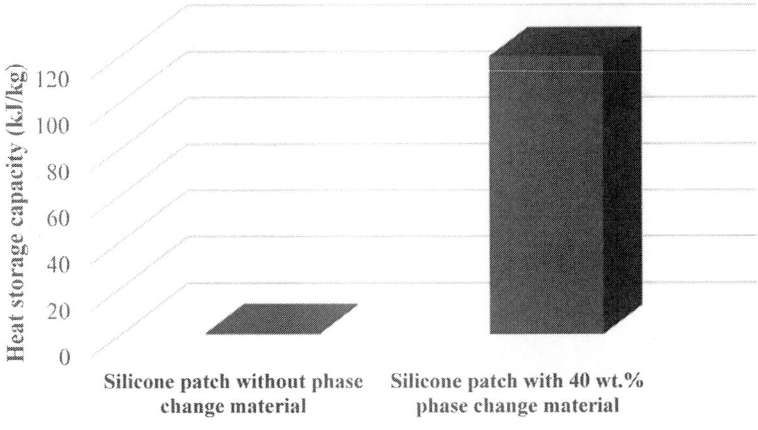

Fig. 5-1: Comparison of the heat storage capacity of silicone patches with and without a phase change material incorporated therein.

After comparing the heat storage capacities of fabrics, polymeric films, and other materials with and without the application of phase change materials, it is obvious that utilizing phase change materials can substantially enhance their heat storage capacity.

5.4. Cooling effect

A cooling effect is obtained by a continuous latent heat absorption of the phase change material. As soon as the temperature of the phase change material reaches its trigger point in a heating process, it starts to absorb latent heat. In this process, the phase change material draws heat away from its surroundings while its temperature remains constant. By taking heat away from adjacent areas, these areas are being cooled. The cooling process lasts as long as the phase change material absorbs latent heat. The cooling effect ends when the latent heat storage capacity of the phase change material is reached. To have a subsequent cooling effect, the phase change material must be recharged by completely releasing the absorbed latent heat.

5.5 Heating effect

When latent heat stored in the phase change material is released into the surrounding area, a heating effect is provided. Before this effect can occur, the phase change material must be charged by absorbing latent heat, preferably until it reaches its full latent heat storage capacity. The phase change material also must be further heated up to a temperature above the heat release trigger temperature. The phase change material temperature remains constant during its release of latent heat. The released latent heat is completely released into the environment. Due to its constant temperature, the phase change material does not contribute to a possible temperature gradient in the surrounding area.

5.6. Thermoregulating effect

The thermoregulating effect established by the phase change material enclosed in a carrier matrix consists of alternating latent heat absorption and latent heat release. This effect can be used to keep the temperature of a surrounding substrate nearly constant. The phase change material acts in this way as a thermostat. The maximum temperature gradient in reference to the ambient temperature shall be considered in designing such a system. Furthermore, it needs to be ensured that a sudden change in the temperature gradient does not occur leading to a temporary increase of the latent heat absorption or latent heat release and destroying the thermal equilibrium the thermoregulating effect is based on. In addition, the phase change material must remain in the phase change temperature range during the entire time of the application.

5.7. Active thermal barrier effect

Looking at a tensile structure or a garment configuration, the heat flux through these systems is determined by the thermal resistance values of their components, which are based on their thermal insulation properties. Thermal insulation properties of materials are affected by their composition, structure (density), and thickness. These insulation values, and hence, the thermal resistances, are constant within certain temperature limitations. Based on these static insulation properties, the heat flux through the systems cannot be regulated in accordance with thermal needs.

Adding phase change material to a membrane fabric, a tensile structure consisting of or to a fabric layer inside a garment configuration, the phase change material creates an active thermal barrier that adapts the system's thermal insulation to prevailing needs. Suppose the temperature of the membrane or the fabric rises and reaches the trigger temperature for the latent heat absorption; the phase change material will start to absorb the excessive heat and will remain at this temperature, preventing a further rise in the heat flux through the system, which would otherwise occur.

On the other hand, if the material temperature drops and reaches the trigger temperature of the latent heat emission, the phase change material will release its stored latent heat and remain at this trigger temperature as long as the latent heat release continues. Therefore, under these circumstances, the heat flux will not rise and will be limited to the intensity of the heat flux that took place before the phase change occurred.

Because this heat flux control feature in both directions is triggered by temperature, materials equipped with phase change material are called "smart materials".

References

1. Pause, Barbara. 2017. "Smart Textiles Created by the Application of Phase Change Material (PCM)". ITMC International Conference on Intelligent Textiles and Mass Customization, Ghent, Belgium, October 15-17, 2017
2. Pause, Barbara. 2010. "Phase change materials and their application in coatings and laminates for textiles". In: Smart textile coatings and laminates, edited by Smith, W.C., 236-250. Woodhead Publishing Ltd.

3. Fleischer, Amy S. 2015. "Thermal energy storage using phase change materials: fundamentals and applications". Springer Briefs in Applied Science and Technology, Thermal Engineering and Applied Science, 2015 Edition, Springer Cham, Heidelberg, New York, Dordrecht, London
4. Kenisarin, M., Mahkamov, K. 2007. "Solar energy storage using phase change materials". Renewable and Sustainable Energy Reviews. No. 11: 1913-1965. doi:10.1016/j.rser.2006.05.005

CHAPTER SIX

MEASUREMENTS OF THE THERMAL EFFECTS

6.1. Overview of the measuring methods

To determine and quantify the thermal effects caused by microencapsulated phase change materials and non-microencapsulated phase change materials in textiles, textile composites, and other sheet-like substrates as well as end-use products, a measuring and evaluation system has been developed and applied [1]. This measuring and evaluation system has been used to test and evaluate more than one thousand textiles and other sheet-like materials treated with phase change material as well as more than one hundred end-use products equipped with the phase change material technology. Furthermore, based on the tests, a material database has been established which will be introduced in Chapter Eight. The system comprises the measuring procedures summarized in Table 6-1.

The test procedures included in Table 6-1 allow for accurate measurements and evaluation of the thermal benefits received from applications of phase change materials in textiles, textile composites, other sheet-like substrates, and various end-use products.

Measuring procedures	Test objects	Measurements
Differential scanning calorimetry (DSC)	Phase change materials, micro-encapsulated phase change materials, textiles, and other sheet-like substrates treated with phase change material	Latent heat storage capacity, phase change temperature ranges
Dynamic heat transfer measurements	Textiles and other sheet-like substrates treated with phase change material	Static thermal resistance of the substrate, dynamic thermal resistance caused by the phase change material

Radiant heat transfer measurements	Textiles and other sheet-like substrates treated with phase change material	Heat flux through the material
Temperature variation tests	Textiles and other sheet-like substrates treated with phase change material	Duration of the phase change effect under various heat inputs and temperature variations
Climatic chamber tests using test models	End-use products treated with phase change material	Microclimate temperature changes over time
Wearing tests carried out by test subjects in climatic chamber test setups and real-world applications	End-use products treated with phase change material	Microclimate temperature changes over time

Table 6-1: Measuring procedures to study the thermal effects.

The test results are mainly used to quantify thermal performance and compare products with and without phase change material treatment. In addition, they are utilized to study influences on the thermal effects discussed in Chapter Seven. The measurements are suitable for product performance evaluation and provide a beneficial research tool.

6.2. Differential scanning calorimetry

Differential scanning calorimetry (DSC) measurements are used to determine the amount of latent heat absorbed by the phase change material during the melting process and the amount of latent heat released by the phase change material during the crystallization process. Furthermore, the temperature ranges in which these phase transitions occur are measured by utilizing this procedure. Averaging the amount of absorbed latent heat and the amount of released latent heat, the latent heat storage capacity of the phase change material is determined. The latent heat storage capacity is given in Joules per gram (J/g). If the phase change material is, for instance, embedded in a textile substrate, the latent heat storage capacity of the textile substrate is given per square meter under consideration of the textile's weight.

The test method involves the heating or cooling of a sample of the test material (phase change material itself, microcapsules including phase change material, or sheet-like substrates with phase change materials incorporated wherein) at a controlled rate in a nitrogen atmosphere through the temperature ranges in which the melting or the crystallization of the phase change material occurs. The heating or cooling rate under which the measurements are carried out is usually 5 K per minute. The difference in heat flow between the test material and a reference material is continuously monitored and recorded over time. A transition, for instance, from solid to liquid is marked by the absorption of energy by the phase change material resulting in a corresponding endothermic peak in the heating curve [2].

Fig.6-1 shows a DSC thermograph received for the latent heat absorption of an octadecane phase change material. The test results indicate that the octadecane phase change material absorbs an amount of latent heat equal to 227 J/g (delta H). The measurement delta H is determined by integrating the area under the heat flow curve. This latent heat is absorbed in a temperature range between 26°C and 33 °C. The maximum latent heat absorption occurs at about 31 °C (melting peak) [3].

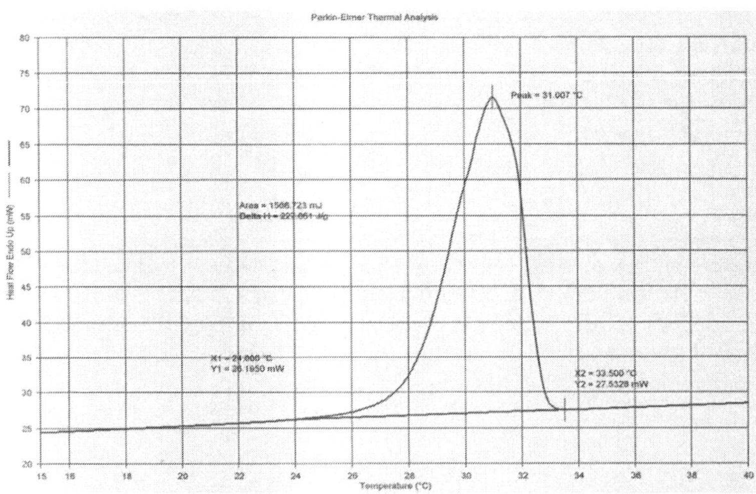

Fig. 6-1: DSC thermograph of the latent heat absorption of octadecane.

On the other hand, a phase transition from the liquid state into the solid state is marked by a release of energy by the phase change material, resulting in an exothermic peak in the cooling curve. This peak in the cooling curve determines the latent heat release and the phase change temperature range in which the latent heat release takes place. Fig. 6-2 shows the DSC thermograph for the latent heat release of the octadecane phase change material.

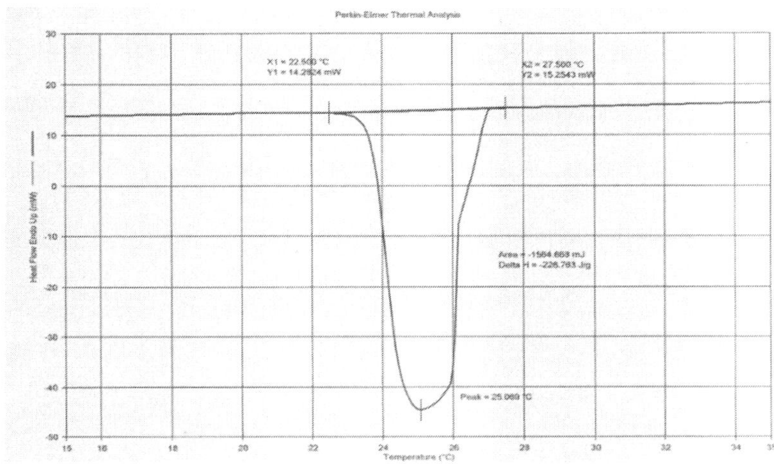

Fig. 6-2: DSC thermograph of the latent heat release of octadecane.

The results shown in Fig.6-2 indicate that the octadecane phase change material releases an amount of latent heat of 227 J/g (delta H). The latent heat is released in a temperature range between 23 °C and 27 °C with a peak crystallization temperature of about 25 °C. The latent heat storage capacity of the octadecane phase change material is then calculated by averaging the values received for the latent heat absorption and the latent heat release. For the selected octadecane phase change material, a latent heat storage capacity of 227 J/g has been calculated [4].

Suppose the phase change material is embedded in a textile structure, the latent heat storage capacity should be calculated in reference to a unit area of one square meter of fabric. This measurement makes a fair comparison of the latent heat storage capacity of fabrics treated with phase change materials possible. For instance, the latent heat storage capacity of a microencapsulated phase change material embedded in a fabric was determined to be 53 J/g, and the overall fabric weight was measured to be

330 g/m². A latent heat storage capacity of 17.5 kJ/m² was calculated for this fabric treated with the microencapsulated phase change material.

6.3. Dynamic heat transfer measurements

Latent heat absorbed or released by a phase change material during its phase transition can be measured by using differential scanning calorimetry (DSC) measurements, as discussed in Chapter 6.2. However, the thermal barrier effect, created by the phase change material when the phase transition occurs in a textile substrate or another sheet-like material, cannot be determined using this procedure. A new measuring and evaluation method was developed whereby the phase change material's latent heat storage capacity is transferred into an insulation term. This method is referred to as "dynamic heat transfer measurement (DHTM)" [5].

The measurement, which represents the thermal barrier function created by the phase change material, is called 'dynamic thermal resistance', and is given in units typically used in thermal resistance measurements. Measuring the dynamic thermal resistance allows for comparison of the thermal performance of, for instance, textiles with and without phase change material treatment. In addition, based on these measurements, the influence of the textile's structure on the thermal barrier effect of the phase change material can be determined. Such information is used to select textile structures that best support the desired thermal benefits of the phase change material. Dynamic thermal resistance values are also utilized in the design process of end-use products with incorporated phase change materials. The test equipment used for the "dynamic heat transfer measurements" is shown in Fig. 6-3.

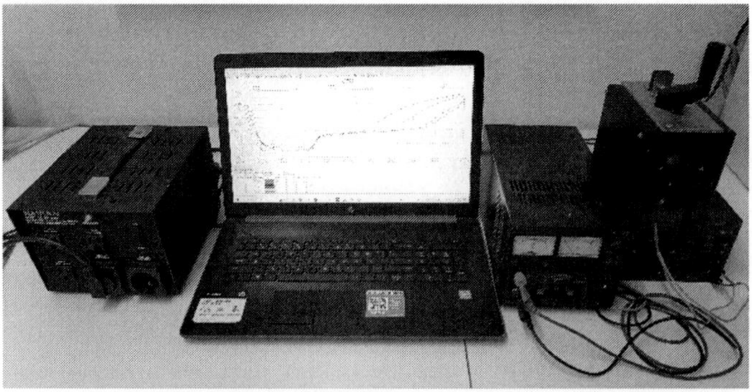

Fig. 6-3: Test equipment used for the "dynamic heat transfer measurements".

In this procedure, the determination of the dynamic thermal resistance provided by the phase change material is separated from the measurement of the static thermal resistance provided by the textile substrate or another sheet-like material [6].

Static thermal resistance of the substrate

To determine the static thermal resistance R_{st} of the substrate without activating the phase transition within the phase change material, a new measuring method was developed that is based on a non-steady state measuring principle. In this method, a short heat impulse is transmitted into the test sample, which increases the sample temperature by only one to two degrees. To prevent the phase transition of the phase change material from taking place, the temperature of the sample must remain outside the phase change temperature range during the entire measurement. Therefore, it is necessary under some circumstances to precool or preheat the test sample.

The measuring procedure consists of two steps. First, the sample is heated from one side until a pre-determined temperature is obtained. Then, the heating process is discontinued, and the sample's temperature drops. The thermal conductivity of the sample material λ is determined from the supplied heat flux ϕ into the sample in the first part of the measuring procedure, as well as the sample's thickness D and temperatures taken at the beginning of the test, the end of the heating period, and the end of the cool-down period which lasts as long as the heating period. δ_1 is the difference of the temperatures measured at the beginning and the end of the heating period. The difference in the temperatures at the end of the heating period and the end of the cool-down period is marked as δ_2.

$$\lambda = \phi\, D / 2\ (\delta_1 - \delta_2) / \delta_1^2 \qquad (6.1)$$

The thermal conductivity is given in W/m K.

The static thermal resistance R_{st} of the substrate is then calculated as the quotient of the substate's thickness D and the substate's thermal conductivity λ according to

$$R_{st} = D/\lambda \qquad (6.2)$$

The static thermal resistance R_{st} of the substrate is given in m^2 K/W [7].

Dynamic thermal resistance generated by the phase change material

Two separate steps are required to determine the dynamic thermal resistance R_{dyn} generated by the phase change material resulting from its absorption or release of latent heat. In preparation for the measurement of the phase change material's barrier function caused by latent heat absorption, sample A, equipped with the phase change material, is pre-cooled so that its temperature is below the phase change temperature range. At the same time, a metal plate is pre-heated so that its temperature is above the phase change temperature range. Then, sample A is brought into contact with the preheated metal plate. The heat emitted from the plate penetrates through the sample and the temperature increase on the other side of the sample is recorded over time. The measurement ends when the temperature on the other side of the sample exceeds that of the range for the phase change, indicating that the phase change is complete. The test is then repeated with a control sample B using the same substrate but without phase change material.

The test setup is shown in Fig. 6-4.

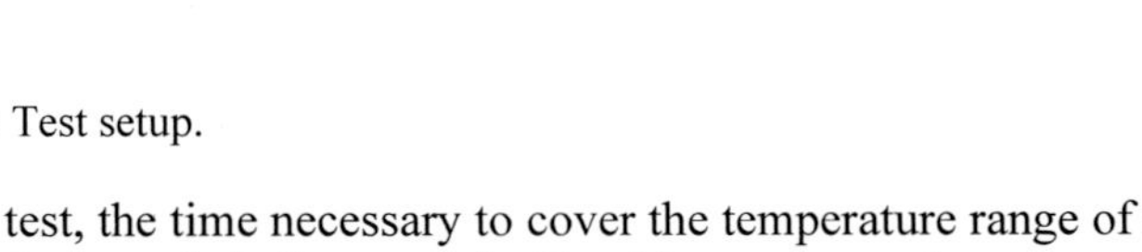

Fig. 6-4: Test setup.

In each test, the time necessary to cover the temperature range of the phase change is recorded. For sample B without phase change material, the measured time T_B is directly related to the static thermal resistance of the substrate R_{stB}. But for sample A equipped with phase change material, the determined time T_A depends on the substrate's static thermal resistance R_{stA} and the dynamic thermal resistance R_{dynA} caused by latent heat absorption of the phase change material.

A temperature versus time plot is shown in Fig.6-5.

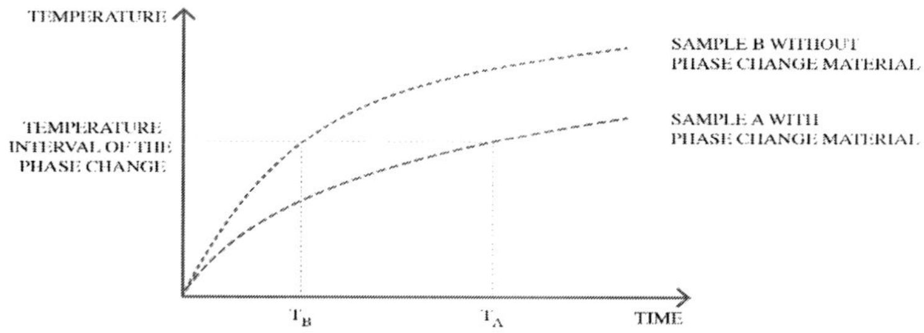

Fig. 6-5: Temperature versus time plot.

Consequently, by comparing the times recorded with reference to the static thermal resistance values, the dynamic thermal resistance R_{dynA} resulting from the latent heat absorption of the phase change material is calculated.

$$R_{dynA} = (R_{stB}\, T_A / T_B) - R_{stA} \tag{6.3}$$

The dynamic thermal resistance R_{dyn} is given in m^2 K/W.

A similar procedure is used to determine the dynamic thermal resistance R_{dyn} of the phase change material due to its latent heat release. In this case, however, the sample is pre-heated, and the plate is pre-cooled.

Total thermal resistance of substrates with and without phase change material

The total thermal resistance R_{total} of a substrate treated with phase change material equals the sum of its substrate's static thermal resistance R_{st} and the dynamic thermal resistance R_{dyn} generated by latent heat absorption or latent heat release of the phase change material.

$$R_{total\ A} = R_{st\ A} + R_{dyn\ A} \tag{6.4}$$

The total thermal resistance R_{total} of a substrate without phase change material equals its static thermal resistance $R_{st.}$

$$R_{total\ B} = R_{st\ B} \tag{6.5}$$

The total thermal resistance R_{total} is given in m^2 K/W.

Test results

The dynamic heat transfer measurements are utilized to quantify the contribution of the phase change material to the textile's or other substrate's thermal performance. The transfer of the latent heat absorption or latent heat release into an insulation term allows for an equitable comparison of materials with and without phase change material treatment. The measuring results are also used to compare and evaluate the heat transfer through substrates that contain different phase change materials or are equipped with the same phase change material in a variety of quantities. Some test results received in tests of fabrics treated with different phase change materials A and B, and A applied in two quantities are shown in Fig. 6-6.

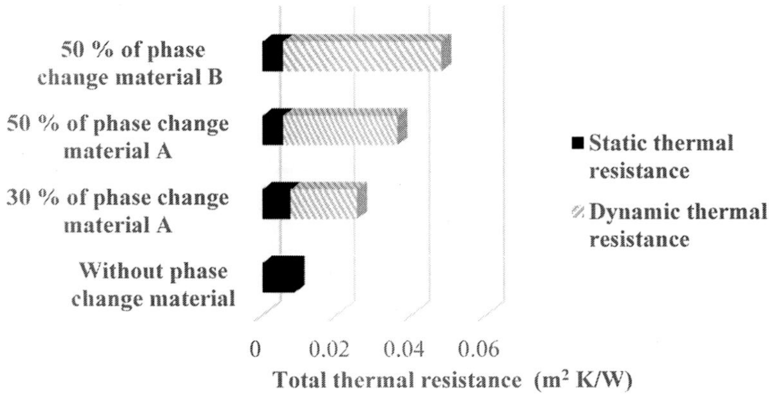

Fig. 6-6: Test results.

In the tests, summarized in Fig. 6-6, a two-millimetre-thick substrate was equipped with two different phase change materials A and B. Phase change material A possesses a latent heat storage capacity of 200 J/g, while phase change material B has a latent heat storage capacity of 280 J/g.

When the phase change material is incorporated into the substrate, the substrate's density increases leading to a decrease in the static thermal resistance of the substrate. The dynamic thermal resistance caused by the phase change material rises with an increase in the phase change material quantity. In addition, the dynamic thermal resistance increases when the phase change material A is substituted by phase change material B which possesses a higher latent heat storage capacity.

Altogether, the test results summarized in Fig. 6-6 indicate a substantial improvement in the substrate's overall thermal insulation when phase change material is applied.

6.4. Radiant heat transfer measurements

The measurements described in Chapter 6.3. are carried out under contact heat exposure. To determine the thermal barrier effect provided by the phase change material incorporated into a substrate such as a roofing material against radiant heat exposure, the test samples are placed in front of a panel heater. The test setup is shown in Fig. 6-7.

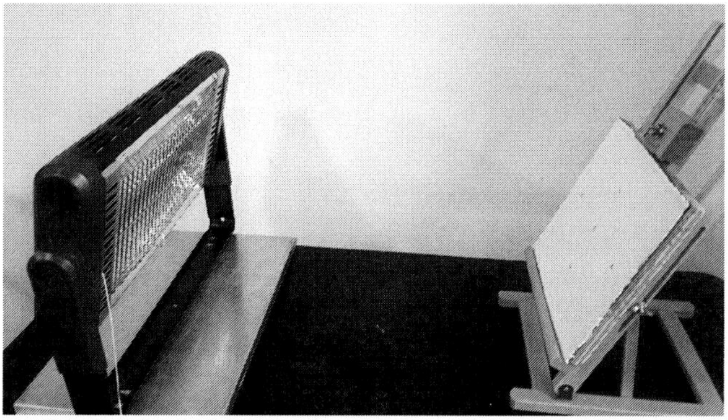

Fig. 6-7: Radiant heat transfer test setup.

The radiant heat provided by the panel heater can be varied to accommodate different heat incidences under certain application conditions. Temperature measurements are taken on the front side and the back side of the test material in given intervals. Based on the temperature measurements, the heat transfer through the test materials is calculated. Results received in a comparison test of roof deck ensembles with and without phase change material treatment are summarized in Fig. 6-8.

The test results summarized in Fig. 6-8 indicate that the latent heat absorption of phase change material reduces the heat transfer through the roof deck ensemble substantially as soon as the trigger temperature of the phase change is obtained.

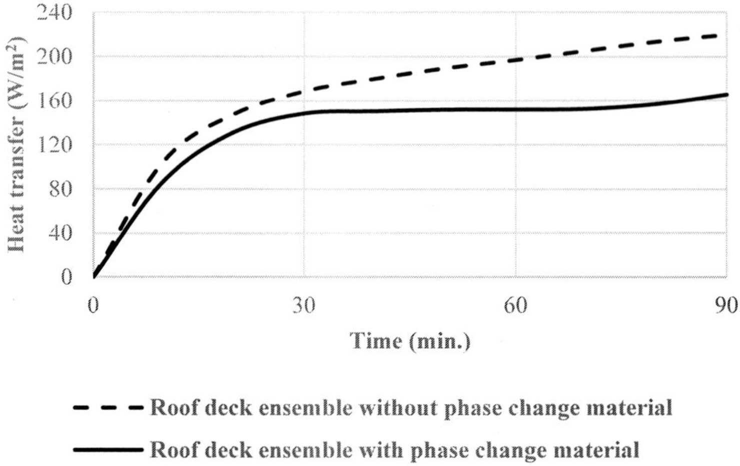

Fig. 6-8: Radiant heat transfer test of roof deck ensembles.

6.5. Temperature variation tests

Temperature variation tests are carried out in a thermal variation testing device. In the tests, the thermal response to temperature variations is observed, and the duration of the thermal effects provided by the phase change material is determined. Furthermore, the thermoregulation effect of the phase change material incorporated into a textile, a textile composite, or another sheet-like material is investigated. The thermoregulating effect provided by the phase change material in a substrate keeps the temperature of the substrate constant in a defined temperature range and for a given duration. The parameters of the thermoregulating effect provided by the phase change material need to be determined under relevant conditions under which the test material is further used.

The thermal variation testing device consists of two chambers, the environmental chamber, and the test chamber. The two chambers are divided by the test material. The environmental chamber can be heated or cooled by a heating/cooling system, and thermocycles over determined time intervals can be programmed into the heating/cooling regiment. By heating or cooling the air in the environmental chamber, a temperature difference arises in reference to the test chamber. A heat flux through the test material between the two chambers is generated resulting from this temperature gradient.

This heat flux through the test material and the temperature variations in both chambers are determined during the test. The temperature changes in the test chamber can quantify the thermoregulating effect of phase change material incorporated into the test material. When the temperature in the test chamber can be stabilized in a temperature range between the melting temperature and the crystallization temperature of the phase change material over desired time intervals, the thermoregulating effect is satisfactory.

Fig. 6-9 shows the test results received for a thermocycling procedure that lasted for one and a half hours. The first part of the test started at a temperature of 20 °C and consisted of a cool-down of the environmental chamber over a temperature range of about 10 degrees Celsius with a cooling rate of one degree per minute. After this cooling procedure, the temperature in the environmental chamber was stabilized at 10 °C for 30 minutes. The test was then continued with the heating cycle where the temperature rose by 30 degrees Celsius with a heating rate of one degree per minute. When a temperature of 40 °C was reached in the environmental chamber, the heating process was stopped, and the chamber temperature was kept constant again for about 30 minutes. Afterwards, the test was continued with another cool-down.

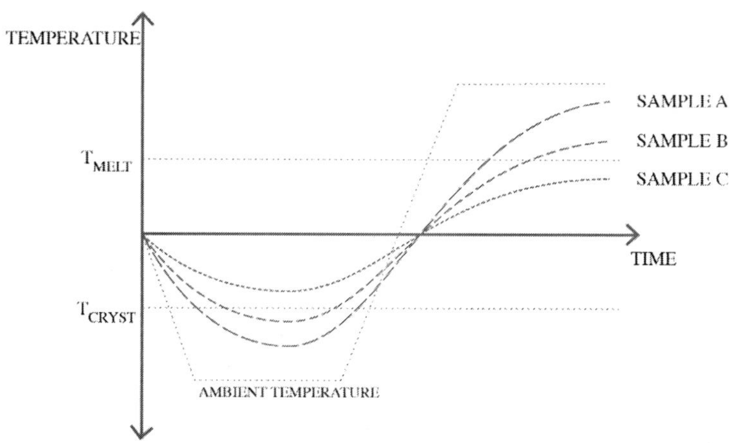

Fig. 6-9: Results of thermocycling tests.

The tests were carried out on three samples of the same substrate equipped with and without phase change material:

- Sample A: without phase change material.
- Sample B: with 85 g/m² phase change material.
- Sample C: with 140 g/m² phase change material.

The thermocycling test results show that when sample A without phase change material was tested, the temperature in the test chamber dropped or rose in accordance with the temperature decrease or increase in the environmental chamber and no temperature stabilization could be reached. On the other hand, using sample C with 140 g/m² phase change material the temperature in the test chamber did not rise above the melting temperature T_{melt} of the phase change material and did not drop below its crystallization temperature T_{cryst} so that a temperature stabilization was obtained. Consequently, a thermoregulating system was established which helps to compensate for undesired temperature fluctuations.

With the lower phase change material content of sample B, temperature stabilization in the given temperature limits (melting and crystallization temperatures) can only be obtained over a short period. Beyond that, the phase change in the phase change material is complete and the further heat flux leads to a temperature decrease or increase corresponding to the temperature change in the environmental chamber. Thus, using sample B also creates a thermoregulating system, but it works under the given conditions for a shorter period compared to sample C.

6.6. Model tests

To test and quantify the thermal performance of end-use products equipped with phase change materials, several test models were developed and implemented. These test models comprise:

- A head form for tests of helmets.
- A hand form for tests of gloves.
- A foot form for tests of sneakers or ski boots.
- A torso form for tests of sleeping bags, comforters, pillows, mattresses, and car seats.

Each test model possesses internal heating and water vapor generating systems. The test models are also equipped with temperature and moisture sensors connected to a logger system. The heating of the forms can be varied to accommodate different wearing situations. The temperatures are measured continuously at different locations on the surfaces of the forms. Water vapor is generated when the overall temperature on the surface of the form rises above a threshold temperature. The moisture content is measured continuously at several locations as well. These sensors are connected to a second logger system. The entire test procedure is computer-controlled. All tests are carried out under a controlled environment inside a climatic chamber.

Fig. 6-10 shows a ski helmet test setup utilizing the head form arranged in a climatic chamber. To carry out the test, the ski helmet is placed on top of the head form. The head form is heated by an internal source. The amount of heat generated and emitted into the microclimate of the helmet is based on the heat emission of a human head while performing certain activities. The provided heat is released uniformly over the head form's back and upper surfaces.

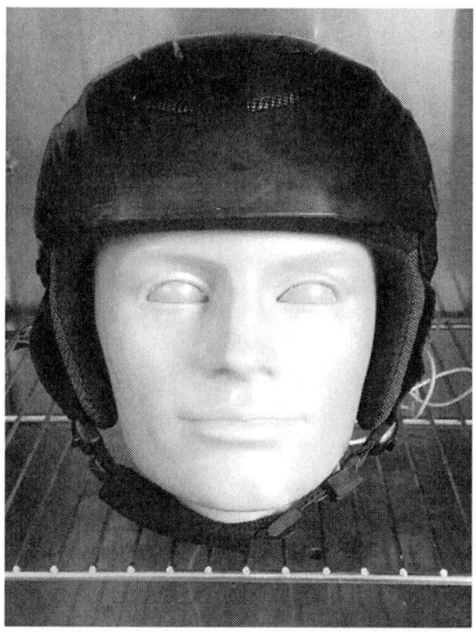

Fig. 6-10: Ski helmet test utilizing the heated and sweating head form.

The water vapor supplied by a humidifier is released through tiny pores in the head form's surface. The amount of water vapor released is linked to the increase in the surface temperature.

The ski helmet was tested with and without phase change material treatment. In one of the test variants, patches equipped with phase change material were inserted into the ski helmet. The melting temperature of the selected phase change material was about 31° C.

The 60-minute test was carried out under a constant heat release by the head form, equal to a heat emission of a human head while cross-country skiing. The climatic chamber temperature was set at 0 °C.

The test results for the microclimate temperature development inside both helmets are shown in Fig. 6-11.

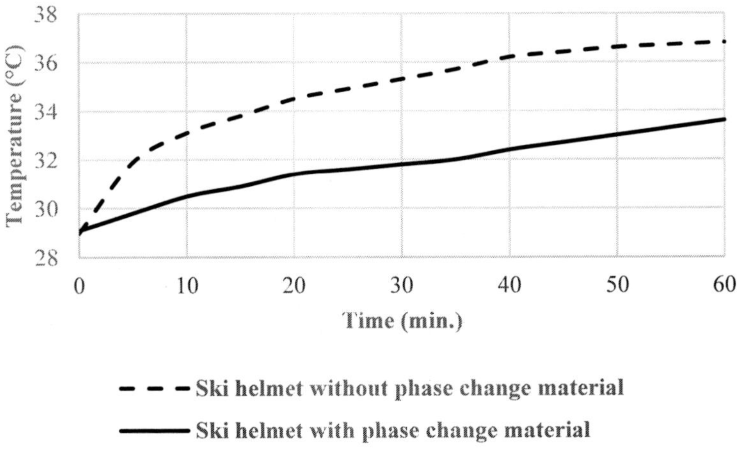

Fig. 6-11: Temperature development in the microclimate inside the ski helmets.

The test results summarized in Fig.6-11 indicate that the phase change material application in the helmet's interior keeps the microclimate temperature in the comfort range below 34°C during the entire test. Without the phase change material application inside the ski helmet, a comfortable microclimate temperature is only maintained for about 20 minutes. The moisture development in the microclimate inside the helmet is shown in Fig. 6-12.

Due to the cooling feature of the phase change material, the moisture buildup in the microclimate is delayed and the moisture content rises only to 50 percent during the entire test period. Without the phase change material application, the moisture content in the microclimate inside the ski helmet exceeded an uncomfortable high amount of 70 percent already after about 30 minutes into the test.

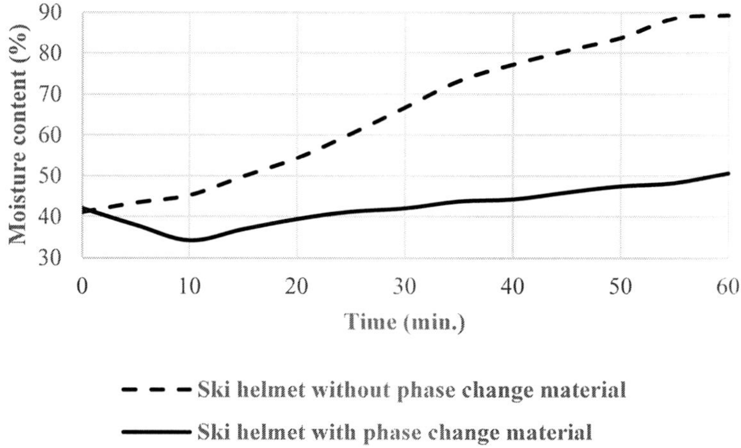

Fig. 6-12: Moisture development in the microclimate inside the ski helmets.

Besides the constant heat output during the entire test, there are often procedures carried out where the heat generation varies in accordance with the alternating activities of a potential wearer.

The climatic chamber temperature can also be altered to accommodate variable ambient temperature conditions. Utilizing the test models eliminates variations in the results which could not be avoided when wearing tests with test subjects are carried out. In addition, these tests also validate the results received in the material tests (Chapter 6.3).

6.7. Wearing tests

Wearing trials with human test subjects are often considered for testing the thermal performance of entire garment systems, where phase change material is incorporated into at least one of the garment layers. In these wearing trials, the test subject performs a specific program of physical activities under given climatic conditions. Such physical activities can be simulated by walking on a treadmill or pedaling on a bicycle ergometer with a defined speed. The tests are performed in a climatic chamber. The chamber's temperature and humidity are either kept constant during the entire test or vary in accordance with a specific test protocol. The human test subjects need to be selected carefully. They must be healthy and meet standard person criteria which means they need to be in a certain range with their body measurements. Because humans are involved in the test, a medical doctor is required to be present and must oversee the entire test. A wearing test of a nonwoven protective garment ensemble carried out on a bicycle ergometer in a climatic chamber is shown in Fig. 6-13.

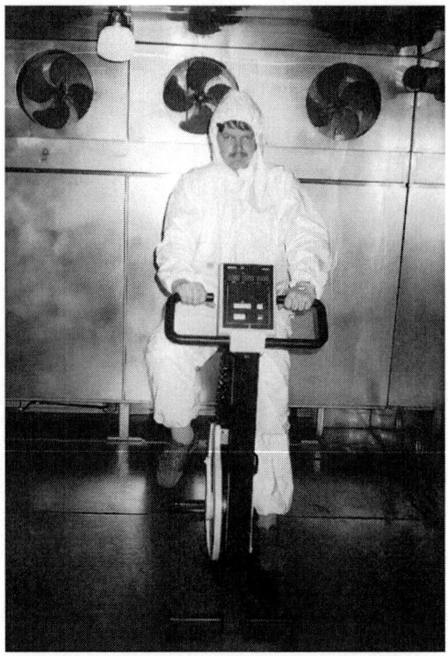

Fig. 6-13: Wearing test of a nonwoven protective garment ensemble carried out on a bicycle ergometer in a climatic chamber.

By utilizing the exercise equipment, the test subject performs the test under one or several activity levels where specific amounts of heat are generated by the body. This heat needs to be released through the garment layers into the environment in a sufficient amount to keep the microclimate temperature in the comfort range. To determine the comfort level the garment system can provide under different wearing situations, temperature and moisture measurements are taken in the microclimate. For these measurements, small temperature and moisture probes are attached to the skin at different locations such as the forehead a, the right scapula b, the left upper chest c, the upper right arm d, the lower left arm e, the left hand f, the right anterior thigh g, and the left calf h. The mean skin temperature t_{sk} is determined in accordance with ASTM Standard F2300-10 by utilizing the following equation:

$$t_{sk} = 0.07\,a + 0.175\,b + 0.175\,c + 0.07\,d + 0.07\,e + 0.05\,f + 0.19g + 0.2\,h \qquad (6.6)$$

To stay within the comfort range and to avoid a rise in the body temperature above 37°C, the mean skin temperature should not rise above 34 °C [8]. In addition, to determine whether the phase change material is staying in the phase change temperature range during the test, it is advisable to attach temperature probes also to the locations where the phase change material is placed. The temperature and moisture development over time during the test is monitored using a logger system connected to a computer.

Besides measuring the temperature and moisture development in the microclimate, heart rates are often monitored to determine the stress resulting from body activities and the wearing of the test garment. Another comfort indication is the sweat production of the human body which is determined by weight measurements of the test subject as well as the single garment parts before and after the test [9, 10].

Test results

In the test, cooling underwear treated with a phase change material was worn underneath an airtight nonwoven protective suit (Fig. 6-13). The thermal performance of this cooling underwear with phase change material was investigated and compared to the thermal performance of regular underwear. The latent heat storage capacity of the phase change material applied to the cooling underwear totalled 48 kJ.

The trials were carried out in a climatic chamber at a temperature of 21°C and a relative humidity of 40 %. During the test, the test subjects rode a bicycle ergometer for 60 minutes without interruption. A metabolic heat rate of about 24 kJ per minute was generated during this activity. 23 kJ per minute was released through the protective garment system by dry heat transfer. During the tests, the skin temperatures and the moisture content were recorded at eight locations on the skin. The mean skin temperature was determined following the equation (6.6). The test results are summarized in Fig. 6-14.

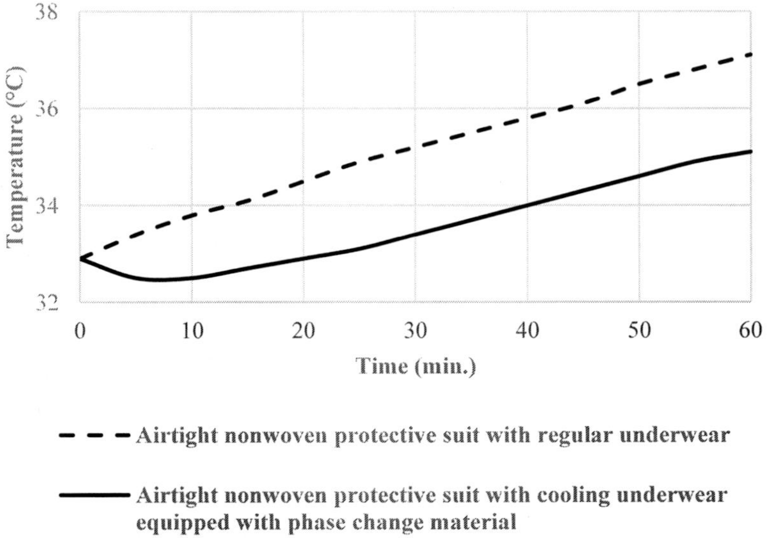

Fig. 6-14: The mean skin temperature development during the wearing test.

The test results summarized in Fig. 6-14 indicate that there is a fast increase in the skin temperature when wearing the airtight nonwoven protective suit with the regular underwear underneath it. After about 45 minutes into the test, the mean skin temperature had already exceeded 36 °C. At this point, a heat stress situation is likely to occur. On the other hand, the cooling effect by the heat absorption of the phase change material starting at about 32 °C leads to a substantial delay in the temperature increase while wearing the airtight nonwoven protective suit with the cooling underwear with phase change material underneath it under the same conditions. At the end of the test, the difference in the mean skin temperature totals about two degrees Celsius. The delay in the temperature rise results in a significantly smaller moisture buildup in the microclimate as shown in Fig. 6-15. The results in Fig. 6-15 are the mean values of the measurements taken at the eight microclimate locations.

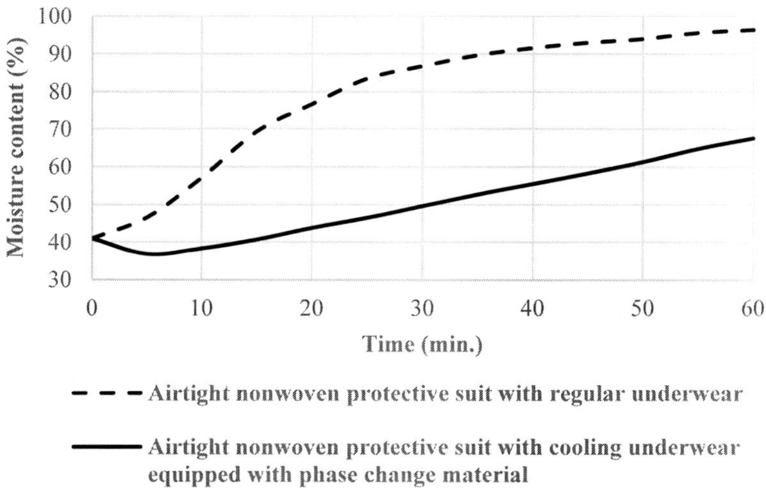

Fig. 6-15: Development of the moisture content in the microclimate during the wearing test.

While wearing the air-tight nonwoven protective suit with regular underwear, the moisture content in the microclimate underneath the suit rises substantially due to the lack of moisture transfer through the material the protective suit consists of. Already after about 15 minutes into the test, the moisture content in the microclimate leads to a feeling of uncomfortable dampness. In contrast, due to the delayed increase in the mean skin temperature by latent heat absorption of the phase change material the cooling underwear is equipped with, a substantially lower amount of moisture is generated by the skin. Therefore, the moisture content in the microclimate is kept at a much lower level throughout the test when the cooling underwear is worn instead of regular underwear underneath the airtight nonwoven protective suit.

The test results received in the climatic chamber tests were verified when the nonwoven protective garment ensemble was worn under real working conditions such as the one shown in Fig. 6-16.

Fig. 6-16: Verification of the climatic chamber test results under real working conditions.

References

1. Pause, Barbara. 2010. "Phase change materials and their application in coatings and laminates for textiles". In: Smart textile coatings and laminates, edited by Smith, W.C., 236-250. Woodhead Publishing Ltd.
2. American Society for Testing and Materials (ASTM) International. 2018. "Standard Test Method for Enthalpies of Fusion and Crystallization by Differential Scanning Calorimetry". Standard E793-06. Book of Standards, Vol.14.01
3. American Society for Testing and Materials (ASTM) International. 2018. "Standard Test Method for Melting and Crystallization Temperatures by Thermal Analysis". Standard E794-06. Book of Standards, Vol. 14.01
4. European Committee for Standardization (CEN). 2016. "Textiles and textile products - Textiles containing phase change materials (PCM) – Part 1: Determination of the heat storage and release capacity". Standard EN 16806-1

5. Pause, Barbara. 2017. "Smart Textiles Created by the Application of Phase Change Material (PCM)". ITMC International Conference on Intelligent Textiles and Mass Customization, Ghent, Belgium, October 15-17, 2017
6. Pause, Barbara. 2015. "Measuring the thermal barrier function of Phase Change Material (PCM) integrated in sheet-like materials." 43rd NATAS Conference, Montreal, Canada, August 11-13, 2015
7. Pause, Barbara Hildegard. 2000. "Procedure and device for the measurement of thermophysical characteristics of textiles and other plate-like materials". Patent No. US 6,116,777
8. American Society for Testing and Materials (ASTM) International. 2022. "Standard Test Method for Measuring the Performance of Personal Cooling Systems using Physiological Testing". Standard F2300-10. Book of Standards, Vol. 11.03
9. Pause, Barbara. 2003. "Nonwoven protective garments with thermo-regulating properties". Journal of Industrial Textiles Vol. 33. No.2: 93-99
10. Pause, Barbara. 2006. "New cooling undergarment for protective garment systems". 3rd ECPC conference, Gdynia, Poland, May 10-12, 2006

CHAPTER SEVEN

INFLUENCES ON THE THERMAL EFFECTS

7.1. Overview of the influences on the thermal effects

There are a variety of parameters, which influence the intensity and, hence, the duration of the thermal effects caused by microencapsulated or non-microencapsulated phase change material.

Such parameters comprise, for instance:

- Quantity of the phase change material.
- Quality of the phase change material.
- Structure of the substrate.

To determine these influences on thermal performance, the investigations have been concentrated on applications of phase change materials in textile substrates [1, 2]. These thermal performance influences must be considered when selecting fabrics with incorporated phase change materials or designing textile end-use products utilizing this technology. The thermal performance influences superimpose each other, which makes it difficult to figure out the rules the thermal effects obtained by the phase change materials in their textile applications follow. To determine these rules, the influences were separated from each other in a defined test series.

The intensity of the thermal effects obtained by the phase change material in a piece of fabric has been quantified using the dynamic heat transfer measurements described in Chapter 6.3. It has been proven that this is the only measuring method, which is suitable, for instance, to quantify the influences of the textile structure on the thermal effects caused by the phase change material incorporated into a fabric. Thus, the following comparisons are based specifically on measurements of dynamic thermal resistance values. The correlation between the latent heat storage capacities and the dynamic thermal resistance values has been validated in the first two studies, in which the influences of the applied phase change material quantity and the phase change material quality on the thermal effects were determined.

Another influence on the thermal effects is in some circumstances the low thermal conductivity of many of the phase change materials. While the low thermal conductivity of the phase change material does not play an important role regarding the thermal performance when they are incorporated into textile substrates, there is a significant influence on the latent heat exchange when they are used for thermal storage in large storage tanks or even smaller containers. However, there are a variety of means available nowadays to enhance the thermal conductivity of a bulk quantity of phase change material which will be discussed in this chapter as well (Chapter 7.7).

7.2 Phase change material quantity

The intensity of the thermal effects of the phase change material depends mainly on the quantity used in a certain product. Test results received for the latent heat storage capacity of the test fabrics and the dynamic thermal resistances by latent heat absorption and latent heat emission of the phase change material under variation of the applied phase change material quantity are summarized in Fig.7-1. In these tests, the same phase change material was applied to the same textile substrate. The phase change material quantities which were utilized varied between 160 g/m^2 and $240 g/m^2$.

An increase in the phase change material quantity applied to a textile substrate leads to a rise in the amount of latent heat that can be absorbed or released by the phase change material and, hence, an increase of the dynamic thermal resistance by latent heat absorption or latent heat emission of the phase change material. The results presented in Fig. 7-1 show a proportional increase of all measurements in accordance with the rise in the phase change material quantity applied to the fabric.

Despite the advantage of the increase in thermal performance, it needs to be considered that the increase in the amount of phase change material added to the substrate leads to a gain in the fabric's weight, which is not always beneficial. Furthermore, an increase in the amount of phase change material utilized in the fabric structure reduces the water vapor transfer through the fabric leading to a compromised breathability of the textile substrate. Therefore, it is necessary to balance thermal performance, weight limitations, and breathability issues in the development process of textiles equipped with phase change material.

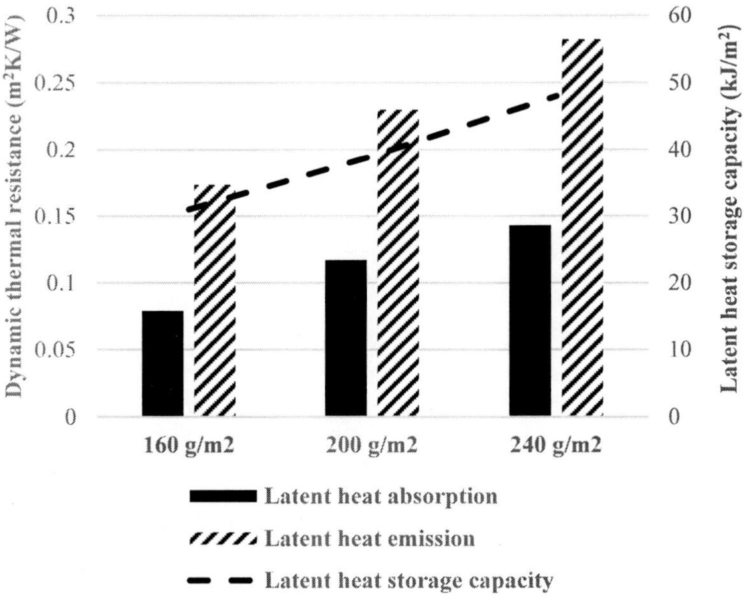

Fig. 7-1: Change in the latent heat storage capacity and the dynamic thermal resistance by latent heat absorption and latent heat emission with a change in the phase change material quantities applied to the textile substrate.

The test results shown in Fig. 7-1 further indicate substantial differences in the dynamic thermal resistance values received for the latent heat absorption and the latent heat emission by the phase change material. Since these different dynamic thermal resistance values are obtained for the same latent heat storage capacity, there need to be other influences on the thermal efficiency of the latent heat absorption and the latent heat emission of the phase change material related to the design of the textile substrate. This has been the subject of further investigations.

7.3 Phase change material quality

Paraffin phase change materials are produced in different purities. These products differ from one another in their latent heat storage capacities. The purer the paraffin-phase change material, the higher its latent heat storage capacity. The purity or quality of the paraffin-phase change material possesses a significant influence on the efficiency of the thermal effect. A comparison of a base paraffin such as octadecane made in different variants is shown in Fig. 7-2.

The variants comprise pure paraffin, technical grade paraffin, and paraffin blend. The test fabric received 160 g/m² of each variant.

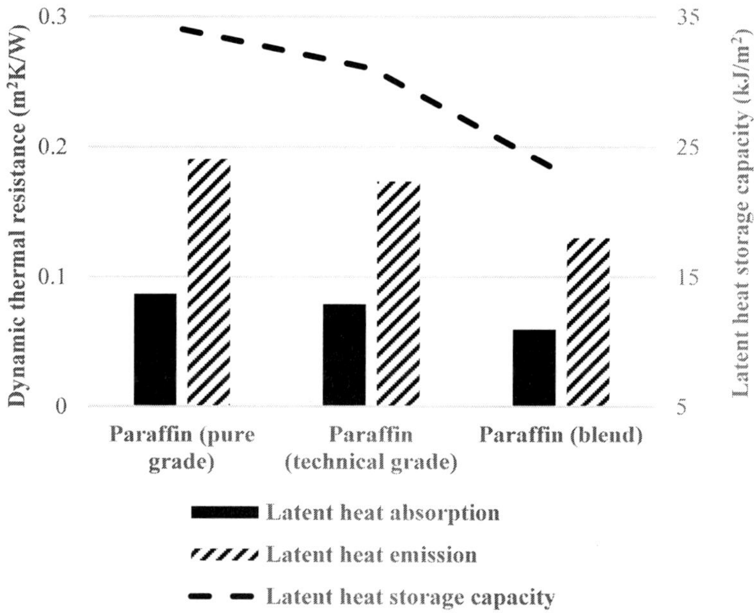

Fig. 7-2: Change in the latent heat storage capacity and the dynamic thermal resistance by latent heat absorption and latent heat emission with a change in the quality of the phase change material applied to the textile substrate.

Utilizing the phase change material variant with the highest quality is advisable to minimize issues such as an undesirable weight addition or reduction in breathability. However, it needs to be considered that an increase in quality leads to a substantial rise in the material price based on the same quantity.

7.4 Structure of the textile substrate

Thickness of the textile substrate

Besides the latent heat storage capacity, which depends on the quality and quantity of the phase change material incorporated into the textile substrate, there is a substantial influence of the substrate's construction on the thermal effects caused by latent heat absorption or latent heat emission of the phase change material. The thickness of the textile substrate is one factor that influences these thermal effects. In a relevant study, the thickness of a textile substrate carrying the phase change material varied between two millimeters and eight millimeters. In these tests, the same phase change material quantity was applied in equal proportions to the test fabrics. The test fabrics were made of the same material. All test fabrics possessed densities of about 130 kg/m². The test results are summarized in Fig. 7-3.

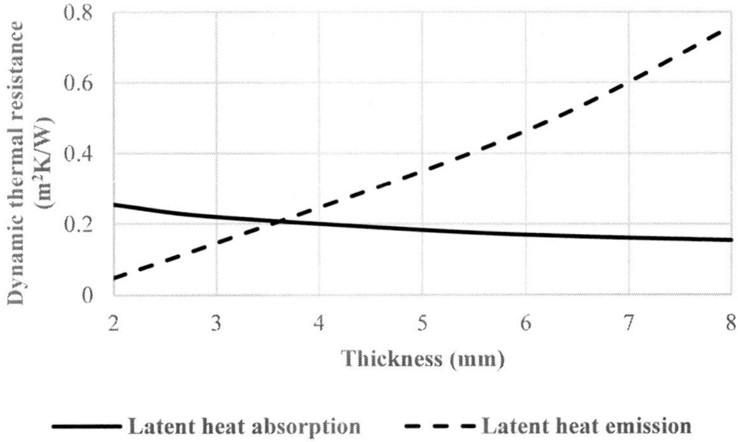

Fig. 7-3: Dynamic thermal resistance by latent heat absorption and latent heat emission of the phase change material in dependence on the substrate's thickness.

The test results summarized in Fig.7-3 indicate that the dynamic thermal resistance by latent heat emission of the phase change material rises substantially when the thickness of the substrate increases. On the other side, there is a slight decrease in the dynamic thermal resistance by latent heat absorption of the phase change material with an increase in the thickness of the substrate.

Thus, the thicker substrates especially support the latent heat emission of the phase change material while thinner substrates mainly support the latent heat absorption of the phase change material. On the other hand, a good thermoregulating effect (equality in the amounts of latent heat absorption and latent heat emission of the phase change material) is obtained when the substrates possess thicknesses between three millimeters and four millimeters.

Density of the textile substrate

The density of the substrate also influences the efficiency of the thermal effects obtained by latent heat absorption or latent heat emission of the phase change material. In a relevant study, the density of the substrate varied between 80 kg/m³ and 140 kg/m³. The thickness of the substrate was kept constant at 2.5 millimeters. The same phase change material quantity was applied to the substrates in equal proportion and equally divided through them. The results received in these tests are shown in Fig. 7-4.

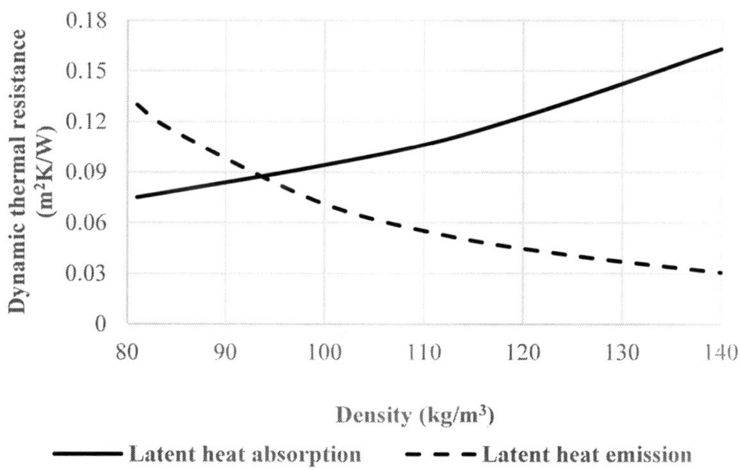

Fig. 7-4: Dynamic thermal resistance by latent heat absorption and latent heat emission of the phase change material in dependence on the substrate's density.

The test results shown in Fig. 7-4 indicate that the dynamic thermal resistance by latent heat absorption of the phase change material increases with a rise in the substrate's density. On the other hand, the dynamic thermal resistance by latent heat emission of the phase change material decreases while the substrate's density increases. Thus, the denser a substrate, the better it supports the latent heat absorption of the phase change material. A sufficient thermoregulating effect can be obtained at a substrate's density of about 90 kg/m^3.

Considering both structural influences of the substrate on the phase change material's thermal effects, the following results are obtained:

- A thin and dense substrate especially supports the latent heat absorption by the phase change material.
- A lofty and thick substrate mainly promotes the latent heat emission of the phase change material.
- A sufficient thermoregulating effect (equal efficiency of the latent heat absorption and latent heat emission of the phase change material) is obtained when the substrate is between three millimeters and four millimeters thick and possesses a density of about 90 kg/m^3.

A dense and thin substrate ensures that the heat applied to the material really reaches the phase change material so that the phase change can take place sufficiently. On the other hand, in a lofty and thick substrate, the heat released by the phase change material stays longer in the substrate's structure, which makes the heat emission by the phase change material very effective.

Compression of the textile substrate

If contact pressure is applied to a fabric, its thickness decreases, and its density rises. Studies have been carried out where fabrics with incorporated phase change material were compressed by a static force. The thickness and density change of the fabric equipped with phase change material in dependence on the applied pressure is shown in Fig. 7-5 and Fig. 7-6.

The fabric used in the tests possesses a weight per unit area of 425 g/m^2 and a basic thickness of 3.2 millimeters. The fabric's compressibility was about 17 percent. A phase change material quantity of 150 g/m^2 was utilized in the fabric. The contact pressure applied to the fabric varied between 0.2 kPa and 5 kPa.

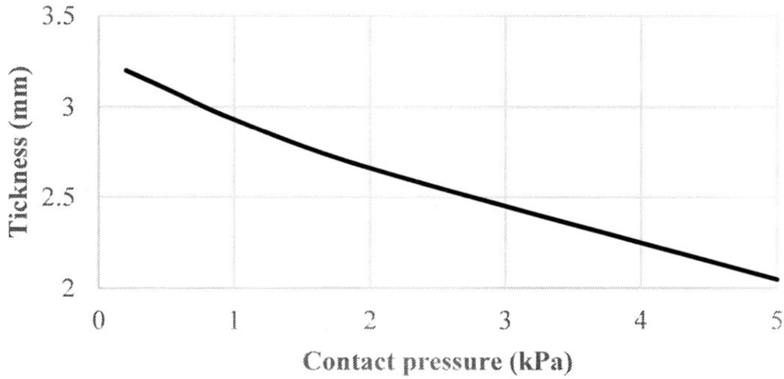

Fig. 7-5: Thickness of the fabric in dependence on the contact pressure.

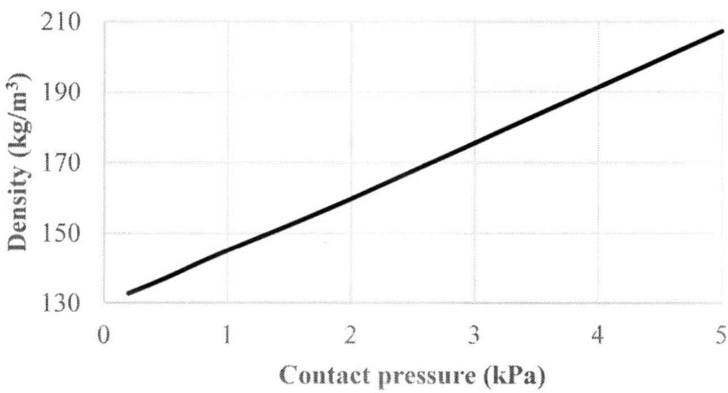

Fig. 7-6: Density of the fabric in dependence on the contact pressure.

Because of the decrease in thickness and the increase in density, increasing the contact pressure applied to the fabric can change the efficiency of the thermal effects. The changes in the static and dynamic thermal resistance values are summarized in Fig. 7-7.

The basic thermal resistance and the dynamic thermal resistance resulting from the phase change material's latent heat emission decrease with a rise in the contact pressure applied to the fabric due to the thickness decrease and the density rise. Thus, the thermal insulation effect caused by the fabric construction and the latent heat emission by the phase change material is reduced when contact pressure is applied to the substrate.

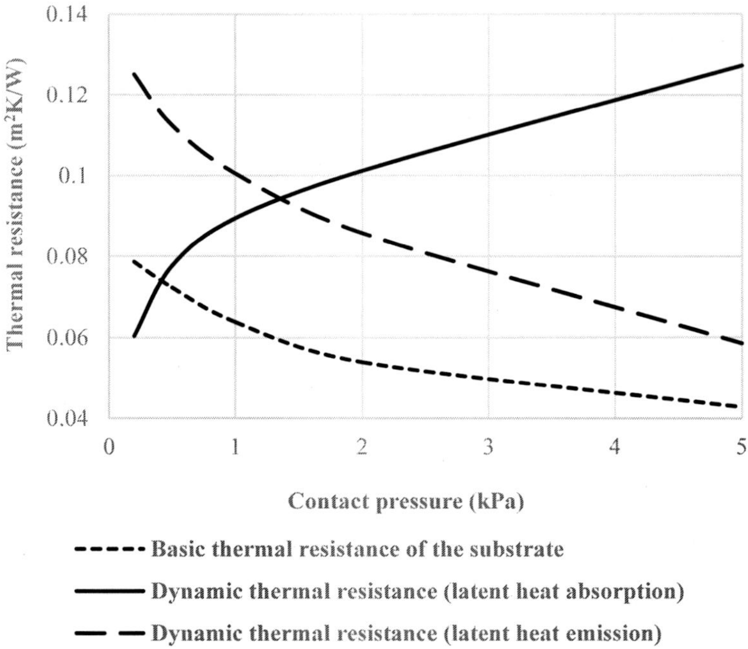

Fig. 7-7: Thermal resistance values in dependence on the contact pressure applied to the fabric.

On the other hand, the dynamic thermal resistance caused by the phase change material's latent heat absorption rises when contact pressure is applied to the substrate. Thus, the efficiency of the latent heat absorption by the phase change material is improved when the textile substrate is compressed.

When a contact pressure is applied to the substrate, the change in the dynamic thermal resistance caused by latent heat absorption or latent heat emission of the phase change material is significantly higher than the change in the substrate's basic thermal resistance. Thus, the dynamic thermal resistance change resulting from an applied pressure needs to be considered when utilizing the phase change material technology.

7.5 Components of the textile substrate

Fiber content

Another study examined the influence of the substrate's basic fiber material on the phase change material's thermal performance. Four woven substrates made of different fibers were used to compare thermal performance. The weights, thicknesses, and densities of the woven substrates were kept as similar as possible.

Furthermore, qualitatively, and quantitatively equal phase change materials were applied to the substrates. The fiber content of the substrates and their thermal conductivities are summarized in Table 7-1.

Fiber	Thermal conductivity, W/m K
Cotton	0.54
Polyester	0.25
Polyamide	0.29
Meta-Aramid	0.13

Table 7-1: Substrate fibers and their thermal conductivities.

The dynamic thermal resistances by latent heat absorption and latent heat emission of the phase change material applied to the four substrates made from the prevailing fiber material are shown in Fig. 7-8.

The results illustrated in Fig. 7-8 indicate that cotton fibers, with their comparatively high thermal conductivity, especially support the latent heat absorption of the phase change material by transporting the heat to the phase change material in a fast and sufficient manner. On the other side, the meta-aramid fiber supports the latent heat absorption and the latent heat emission of the phase change material in nearly equal proportion.

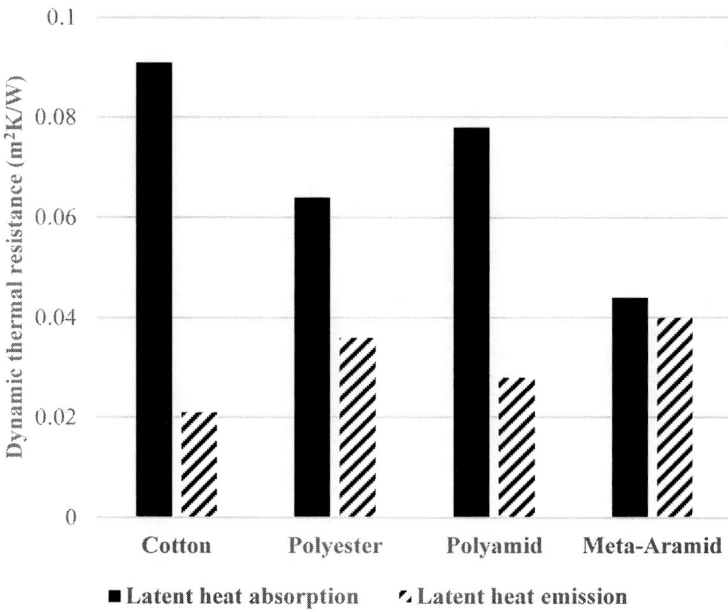

Fig. 7-8: Dynamic thermal resistance by latent heat absorption and latent heat emission of the phase change material received for fabrics made of different fibers.

Moisture content

A fabric treated with phase change material was used to determine the influence of the fabric's moisture content on the thermal performance of the phase change material. The test fabric was conditioned under different relative humidities in a climatic chamber which led to changes in the fabric's moisture content. The increase in the moisture content of the fabric with the rise in relative humidity is shown in Fig. 7-9.

The increase in the fabric's moisture content leads to a slight increase in its thickness and a small decrease in its density. Because of the increased thickness and decreased density of the fabric, an increase in the latent heat emission of the phase change material and a decrease in the latent heat absorption of the phase change material would be expected.

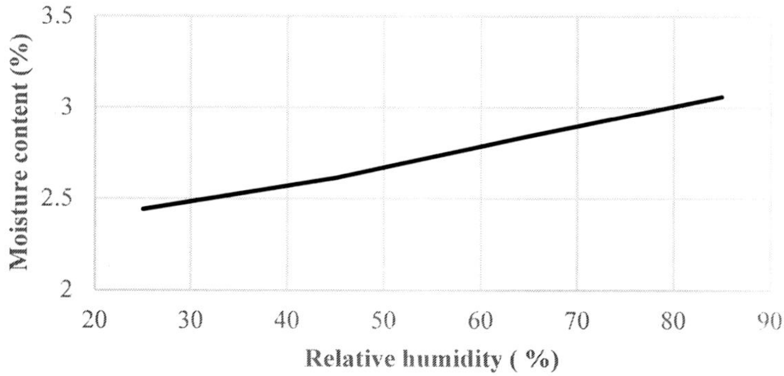

Fig. 7-9: Moisture content of the test fabric in dependence on the relative humidity.

However, the increase in the heat transfer through the fabric because of the increase in the fabric's moisture content acts in opposition to the other two effects. In combination with all effects, the latent heat absorption and latent heat emission of the phase change material remain nearly constant under variation of the substrate's moisture content as is shown in Fig. 7-10.

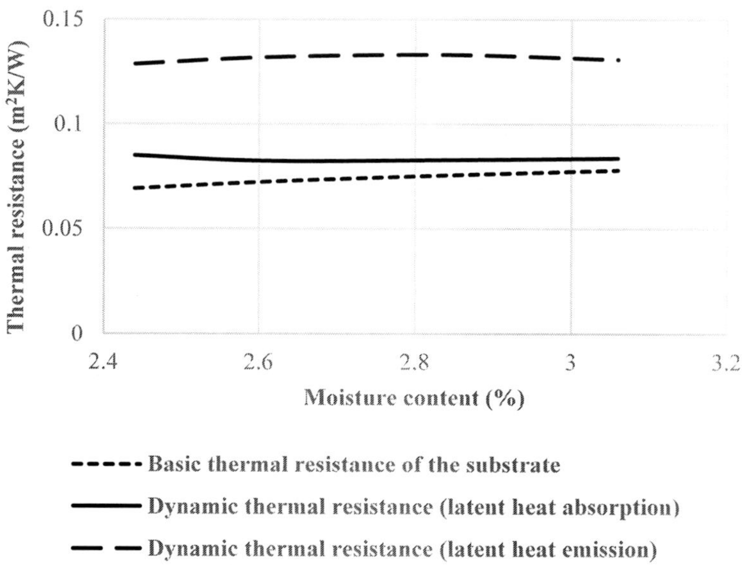

Fig. 7-10: Thermal resistance in dependence on the fabric's moisture content.

Finishes of the substrate

Fabrics are often treated with finishes such as flame-retardant agents to prevent the propagation of fire. Incorporating a bacteriostatic agent will inhibit the growth of bacteria, while a biocidal antimicrobial agent will kill bacteria or fungi. A study was carried out to determine if such finishes influence the thermal performance of the phase change material.

The tests have shown that there is no significant change in the phase change material's thermal performance if the additional treatment does not change the content of the phase change material in the substrate. This is true at least for antimicrobial treatment. For this purpose, the phase change material's thermal performance incorporated into a fabric with and without an antimicrobial treatment was tested. The test results indicate that the antimicrobial treatment slightly extends the phase change temperature ranges, as shown in Fig. 7-11.

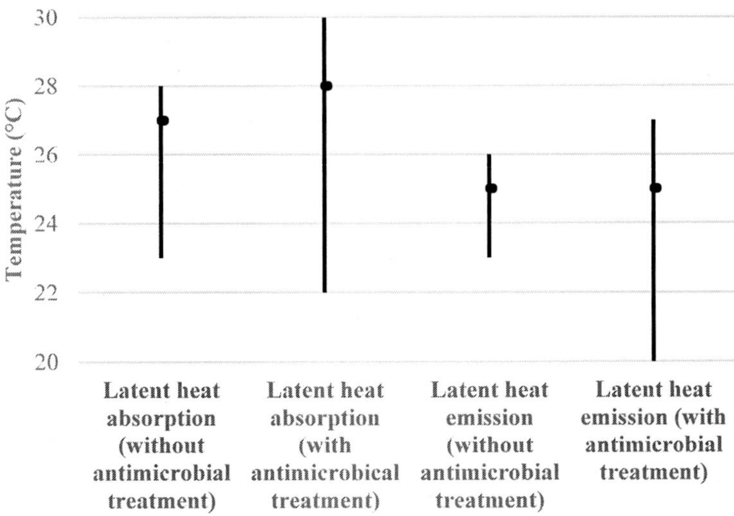

• Marks the peak melting (crystallization) temperature

Fig. 7-11: Phase change temperature ranges with and without antimicrobial treatment of the substrate.

However, the dynamic thermal resistances by latent heat absorption and latent heat emission have not been influenced by the antimicrobial treatment as indicated in Fig. 7-12.

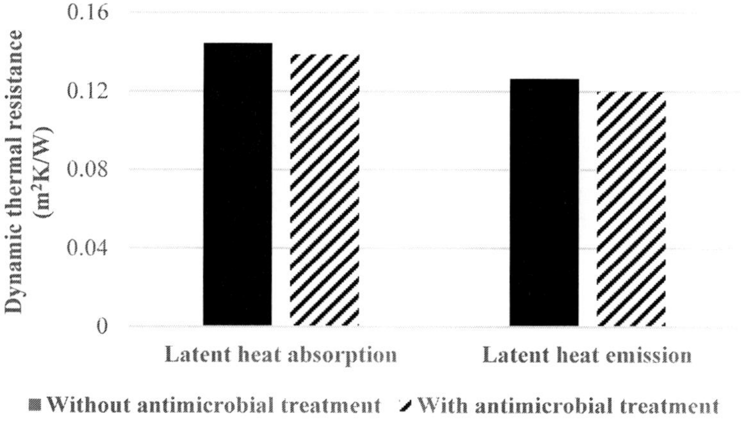

Fig. 7-12: Dynamic thermal resistance by latent heat absorption and latent heat emission of the phase change material with and without antimicrobial treatment of the substrate.

Washing of the substrate

Garment's wear and tear makes it necessary to wash or clean them regularly. To determine if there is, for instance, a loss in microencapsulated phase change material and, hence, a reduction in thermal performance because of washing and cleaning procedures, repeated washing cycles have been carried out. The phase change material's thermal performance was tested after up to forty washing cycles had been performed.

The tests have shown that there is only a significant loss in micro-encapsulated phase change material if microcapsules are located on the fabric's surface where they are unprotected. Therefore, it is recommended to lock the microcapsules into a textile system, such as fibers or foams. In the case of coating the microencapsulated phase change material onto the surface of a textile structure, this surface should be covered with a second coating to prevent a loss of microcapsules out of the first coating layer during washing or cleaning cycles. Fig. 7-13 shows thermal performance test results for a fabric with microencapsulated phase change material after up to forty washing cycles.

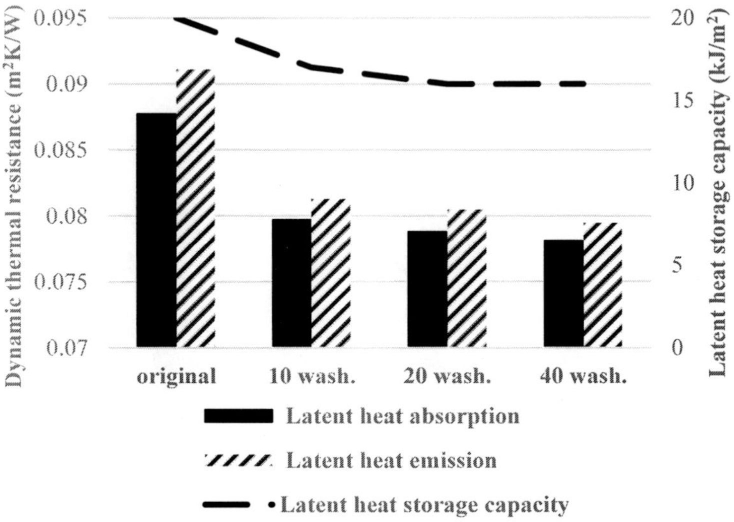

Fig. 7-13: Thermal performance of the microencapsulated phase change material after up to forty washing cycles.

The test results indicate that there is a loss in latent heat storage capacity of about 15 percent within the first ten washing cycles, probably due to a loss in microencapsulated phase change material. Within the next ten washing cycles, the loss in microencapsulated phase change material and, hence, the loss in latent heat storage capacity is less than 5 percent. Between twenty and forty washing cycles, the microcapsule content and, thus, the latent heat storage capacity remains constant.

The dynamic thermal resistance by latent heat absorption and latent heat emission of the phase change material decreases mainly within the first ten washing cycles due to the loss in latent heat storage capacity and possible changes in the fabric construction resulting from the washing process.

7.6 Textile composites

The thermal performance of the phase change material incorporated into a textile composite is influenced by the heat transfer through adjacent layers the composite consists of. The composite structures shown in Fig. 7-14 are used to explain the influence of the multilayer composite constructions on the phase change material's thermal performance.

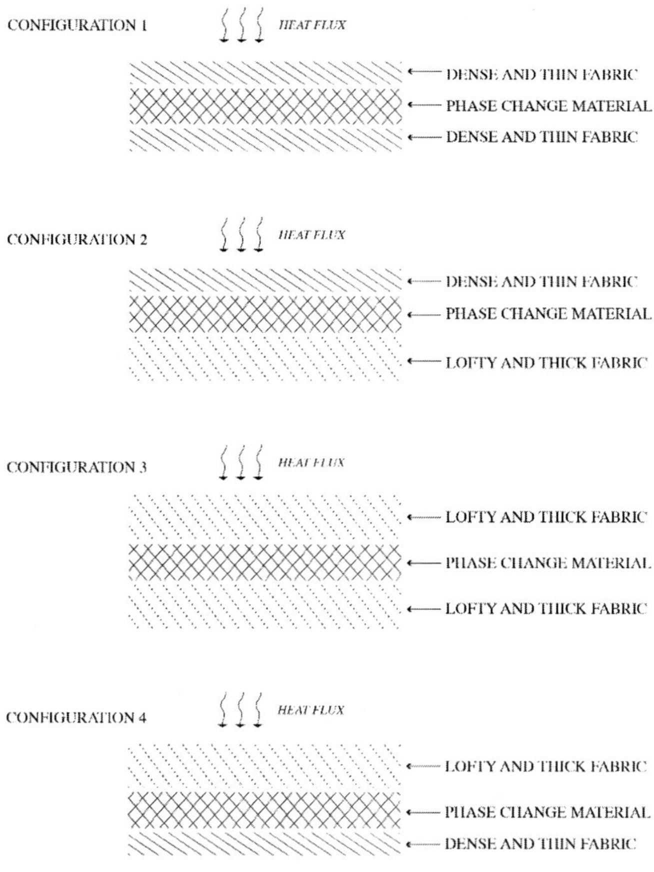

Fig. 7-14: Composite variants.

In the sample configurations, qualitatively and quantitatively equal phase change materials were applied. For the composite configurations summarized in Fig.7-14, the dynamic thermal resistance by latent heat absorption and latent heat emission of the phase change material was determined. The test results are shown in Fig. 7-15.

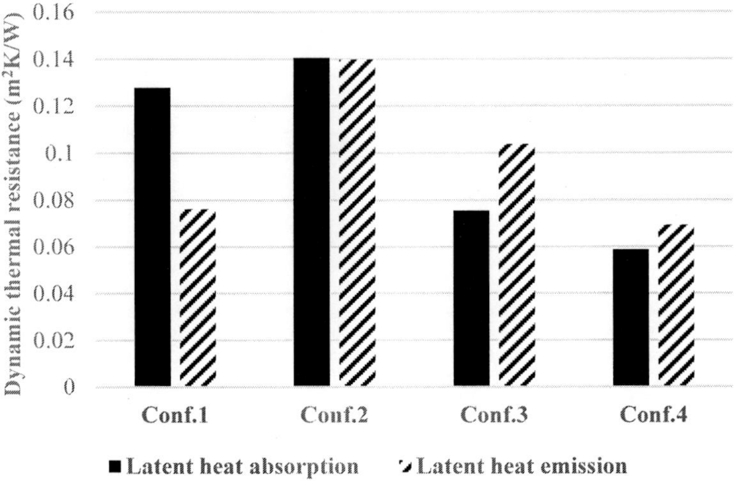

Fig. 7-.15: Thermal performance of the phase change material in dependence on the heat transfer through adjacent layers.

The results shown in Figure 7-15 indicate:

CONFIGURATION 1:

The thin and dense fabric in front of the phase change material layer allows for good heat absorption by the phase change material; however, due to the thin and dense fabric on the back side of the phase change material layer, the heat release takes place inefficiently fast. Thus, the dynamic thermal resistance value by latent heat emission of the phase change material is comparatively low.

CONFIGURATION 2:

The second configuration supports the thermal performance of the phase change material in the best possible way. The thin and dense fabric layer in front of the phase change material layer provides a sufficient heat transfer to the phase change material. The result is a high dynamic thermal resistance value by latent heat absorption of the phase change material. On the other hand, the lofty and thick fabric on the back side of the phase change material layer supports the heat release of the phase change material in a highly efficient way. Therefore, the dynamic thermal resistance value by latent heat emission of the phase change material is also extraordinarily high.

CONFIGURATION 3:

In this configuration, the lofty and thick fabric layer in front of the phase change material layer prevents sufficient latent heat absorption. Thus, there is only a low dynamic thermal resistance value by latent heat absorption of the phase change material. Despite this disadvantage, the latent heat emission of the phase change material is more efficient because the absorbed latent heat can stay longer in the lofty and thick fabric on the back side of the phase change material layer.

CONFIGURATION 4:

This fabric construction leads to the lowest thermal performance of the phase change material in comparison of all four configurations. The lofty and thick fabric in front of the phase change material layer delays the heat transfer to the phase change material layer and makes the latent heat absorption by the phase change material less efficient. On the other hand, there is a very fast and inefficient latent heat emission of the phase change material through the thin and dense fabric layer behind the phase change material layer. As a result of this composite construction, the dynamic thermal resistance values by latent heat absorption and latent heat emission of the phase change material are comparatively low.

7.7. Thermal conductivity enhancement

In thermal energy storage systems, the thermal energy is transferred from an outside source to the phase change material inside the storage system. To store the thermal energy properly, the transition of the phase change material from its solid state into its liquid state must occur in the required time frame. Specifically, organic phase change materials possess very low thermal conductivities and, therefore, provide a high thermal resistance against the transfer of thermal energy. In addition, the thermal conductivity of the phase change material is lower in its liquid state than in its solid state. While the phase change material melts from a solid into a liquid, the heat transfer slows down. The high thermal resistance decreases the thermal performance of thermal energy storage systems. To eliminate this problem, various studies were carried out to find sufficient thermal conductivity enhancements [3].

The following means were investigated and used in thermal energy storage systems:

- Extended surfaces through fins and heat pipes.
- Incorporation of highly conductive fillers such as
  - Graphite,
  - Metal foam composites,
  - Nanomaterials.

Extended surfaces through fins and heat pipes

The most widely investigated thermal conductivity enhancement method is the technique of extended surfaces. In this technique, the surface between the heat source and the heat storage medium, i.e., the phase change material is increased by installing fins and heat pipes inside the container where the phase change material is stored. The extended surface substantially increases the heat transfer to and from the phase change material substantially. However, installing such surface enhancing means adds complexity to the system, increasing the overall costs and decreasing the phase change material quantity inside the container, which reduces the latent heat storage capacity [3].

Incorporation of highly conductive fillers

Graphite

Expanded graphite, an acid-treated graphite, was utilized in a study carried out by Choi et al. to enhance the thermal conductivity of erythritol, paraffin, and stearic acid. Expanded graphite is lightweight and possesses high thermal conductivity, stability, and low density. For instance, when 20wt.% of expanded graphite was applied to erythritol, the thermal conductivity increased from 0.745 W/m K to 1.371 W/m K [4].

Ding et al. blended polyethylene glycol with 10 wt.% of expanded graphite. As a result, the thermal conductivity was enhanced from 0.2985W/m K obtained for pure polyethylene glycol in the solid state to 1.3240 W/m K measured for the blend [5,6].

Metal foam composites

Highly conductive porous materials, such as metallic foams, are used to enhance the heat transfer rate in bulk phase change materials. For instance, Metallic foams are made of copper, nickel, or aluminum and are available in different porosities and pore densities.

The phase change materials are embedded in the porous foams. In this arrangement, the surface area for the heat transfer to and from the phase change material is increased. The increase in the surface area and the high thermal conductivities of the metals lead to a substantial increase in the heat transfer rate between three and ten times. The heat transfer rate is further increased with smaller pore sizes due to a rise in the surface area. However, the density of the pores affects the heat transfer rate only slightly [7].

Nanomaterials

Nanomaterials such as carbon and graphite nanofibers, carbon nanotubes, carbon nanohorns, graphite nanoplatelets, and graphene nanosheets have also been used to enhance the thermal conductivity of phase change materials. Carbon-based nanostructures have similar densities to the organic phase change materials, which results in sufficient stability in the phase change material matrix. In addition, the carbon-based nanostructures possess a very high thermal conductivity, which makes them especially suitable for heat transfer enhancements.

Harish et al. studied the thermal conductivity enhancement in a lauric acid phase change material by applying graphene nanosheets, carbon nanotubes, and carbon nanohorns. They found that the thermal conductivity enhancement was significantly higher when the lauric acid phase change material was in the solid state compared to its liquid state. During the solidification process, the lauric acid forms needle-like structures. The nanoparticles get between these needles, creating a conglomeration of nanoparticles and increasing thermal transport. The test results further indicate that the maximum thermal conductivity enhancement was obtained in the presence of graphene nanosheets as the nanofiller candidate, followed by carbon nanotubes and carbon nanohorns [8].

Abdellatif et al. studied the effects of inserting carbon nanotubes into liquid paraffin wax in a quantity of less than 1 wt.% on the latent heat storage capacity and the thermal conductivity of the paraffin wax matrix. The results showed that a high increase in thermal conductivity was already achieved by this relatively small amount of carbon nanotubes. Further studies showed that the latent heat storage capacity increased linearly with the amount of carbon nanotubes added to the system [9].

In contrast, the results received in the study of Fleischer et al. showed a significant decrease in the latent heat storage capacity of paraffin phase change material by adding herringbone-style graphite nanofibers to the matrix in a concentration of up to 11 wt.%. For instance, the latent heat absorption decreased from 272 J/g without the nanofibers to 243 J/g with an 11 wt.% addition of the nanofibers [10].

References

1. Pause, Barbara. 2017. "Smart Textiles Created by the Application of Phase Change Material (PCM)". ITMC International Conference on Intelligent Textiles and Mass Customization, Ghent, Belgium, October 15-17, 2017
2. Pause, Barbara. 2010. "Phase change materials and their application in coatings and laminates for textiles". In: Smart textile coatings and laminates, edited by Smith, W.C., 236-250. Woodhead Publishing Ltd.
3. Cabeza, Luisa F., Gasia, Jaume, Miro, Laia. 2016. "Materials and system requirements of high temperature thermal energy storage systems: A review. Part 2: Thermal conductivity enhancement techniques". Renewable and Sustainable Energy Reviews, No. 60: 1584-1601. DOI: 10.1016/j.rser.2016.03.019
4. Choi, Bo-Kyung, Ko, Jae-Kyoung, Park, Soo-Jin, Seo, Min-Kang. 2016. "Influences of expanded graphite on thermal characterizations of erythritol, paraffin, and stearic acid for low-temperature phase changer material composites". International Journal of Engineering Science & Research Technology: 796-807. DOI: 10.5281/zenodo.155378
5. Ding, Jing, Fang, Yutang, Yan, Jinyue, Yang, Xiaoxi, et al. 2009. "Preparation and thermal properties of polyethylene glycol/expanded graphite blends for energy storage". Applied Energy, Vol. 86: 1479-1483. DOI: 10.1016/j.apenergy.2008.12.004
6. Gluesenkamp, Kyle R., Graham, Samuel, Hirschey, Jason, Kumar, Navin, et al. 2021. "Thermal Charging Rate of Composite Wax-Expanded Graphite Phase Change Materials". 6th International High Performance Buildings Conference at Purdue, USA, May 24-28, 2021
7. Ali, Hafiz Muhammad, Janjua, Muhammad Mansoor, Rehman, Tauseef-ur, Saijad, Uzair, et al. 2019. "A critical review on heat transfer augmentation of phase change materials embedded with porous materials/foams". International Journal of Heat and Mass Transfer, Vol.135: 649-673. https://doi.org/10.1016/j.jheatmasstransfer.2019.02.001

8. Harish, Sivasankaran, Kohno, Masamichi, Orejon, Daniel, Takata, Yasuyuki. 2017. "Enhanced thermal conductivity of phase change nanocomposite in solid and liquid state with various carbon nano inclusions". Applied Thermal Engineering. Vol. 114: 1240-1246. http://dx.doi.org/10.1016/j.applthermaleng.2015.01.056
9. Abdellatif, O.E., Alshaer, W.G., Barrio, Palomo del, Nada, S.A., et al. 2013. "Analysis of the anomalous thermal properties of phase change materials based on paraffin wax and multi walls carbon nanotubes". International Journal of Heat and Mass Transfer, Vol.1. No.5: 297-307
10. Fleischer, Amy S., Warzoha, Ronald J., Weigand, Rebecca M. 2015. "Temperature-dependent thermal properties of a paraffin phase change material embedded with herringbone style graphite nanofibers". Applied Energy, Vol.137: 716-725 http://dx.doi.org/10.1016/j.apenergy.2014.03.091

CHAPTER EIGHT

DESIGN OF END-USE PRODUCTS EQUIPPED WITH PHASE CHANGE MATERIAL

8.1. Basic principles of the product design

Applying the right phase change material in a proper quantity to the right location in a product configuration is very important for the successful application of the phase change material technology. In the past, many products utilizing phase change material technology failed in tests because these basic principles were not met.

Selecting the right phase change material means that its phase change temperature range needs to be in accordance with the product's application temperature range. Otherwise, none or only a partial phase change effect will occur, and the desired thermal benefits will not be achieved. To reach the intended duration of the thermal effects, a sufficient quantity of phase change material must be applied.

Furthermore, a substrate needs to be selected that supports the thermal effects generated by the phase change material application as much as possible. Other factors, such as weight and thickness restrictions or requirements regarding breathability and mechanical stability of the substrate, must also be considered in the product design. Finally, the substrate treated with the phase change material needs to be arranged in the right location of the end-use product to be thermally effective.

This chapter explains the product design process using a garment ensemble. However, this design process is also valid for the development of all kinds of other applications utilizing phase change material [1, 2, 3].

8.2 Design procedure

To support the complex design process of end-use products equipped with phase change material, a computer-aided calculation and simulation procedure has been developed. This procedure facilitates the designer's work and prevents requirements for repeated testing. The computer-aided development also helps to avoid faulty designs where the desired thermal effects do not occur expectedly or do not appear at all. Before the design process can be started, the current product to which the phase change material shall be applied needs to be subjected to a preliminary examination.

Following this preliminary examination, the design procedure consists of the following steps:

- Phase change material selection.
- Estimation of the sufficient phase change material quantity.
- Selection of a suitable substrate.
- Design of the end-use product.

8.3 Stages of the design process

The different stages of the design process are illustrated by utilizing the phase change material application in a ski wear ensemble.

8.3.1 Preliminary examination of the existing product

The selected ski wear ensemble includes a ski suit made of an inner liner, an insulation layer, and a shell layer. The ski suit comprises a jacket and pants. Thermal underwear and a fleece sweater are worn underneath the ski suit.

Wearing this garment ensemble, the thermal insulation of the garment layers limits the heat transfer from the human body into the environment. This passive thermal insulation effect can only be adjusted to the often changing thermal wearing conditions during the various skiing activities by adding or removing garment layers. But this is not always possible. Thus, strenuous activities during downhill skiing or cross-country skiing can often lead to overheating the wearer's body and, hence, increased sweat production (Fig. 8-1).

Fig. 8-1: Downhill skiing.

In these thermal stress situations, the heat generated by the body exceeds the heat released through the garment layers into the environment in a timely manner.

In contrast, sitting on the ski lift or standing and waiting at the lift at low ambient temperatures frequently results in a decrease in skin temperature. (Fig.8-2). The heat loss through the garment layers, which exceeds the body's heat production, is responsible for the decrease in the skin temperature and, consequently, the body temperature.

Applied to a ski wear ensemble, the phase change material would generate an active thermal barrier effect adapting the ensemble's thermal insulation to the different wearing conditions. This leads to a substantial improvement in the ensemble's thermo-physiological wearing comfort.

In this preliminary stage of the development process, it is necessary to determine the desired composition of the product, select the product components, measure their weights and thicknesses, and investigate the dry and wet heat transfer through them.

Fig. 8-2: Sitting on a ski lift.

The ski wear ensemble consists of the following components:

- Thermal underwear (100 percent wool).
- Fleece sweater (100 percent polyester).
- Ski suit:
    - Inner liner (100 percent polyamide 6.6),
    - Insulation batting (100 percent polyester),
    - Shell layer (100 percent polyamide).

After selecting the components of the ski wear ensemble, their weights and thicknesses have been determined in accordance with ASTM Standard D3776/D3776M-20 and ASTM Standard D1777-96 [4, 5].

The weight and thickness of the components creating the ski wear ensemble are summarized in Table 8-1.

Ski wear ensemble	Weight per unit area, g/m²	Thickness, mm
Thermal underwear	195	1.3
Fleece sweater	275	4.6
Inner liner	80	0.2
Insulation batting	75	3.3
Shell layer	155	0.3

Table 8-1: Weight and thickness of the components of the ski wear ensemble.

The dry and wet heat transfer through these fabrics is determined by measuring their thermal and evaporative resistance values following ASTM Standard F1868-23 [6]. The test results are summarized in Table 8-2.

Ski wear ensemble	Thermal resistance, m^2 K/W	Evaporative resistance, m^2Pa/W
Thermal underwear	0.0236	3.5
Fleece sweater	0.1278	5.0
Inner liner	0.0017	4.1
Insulation batting	0.1189	3.3
Shell layer	0.0021	5.9

Table 8-2: Thermal and evaporative resistance values of the ski wear ensemble.

Besides these measurements, the conditions under which the product, i.e., the ski wear ensemble, is worn need to be determined in these preliminary investigations as well. The conditions comprise, for instance, physical activities and their possible durations, as well as ambient temperatures. After concluding the preliminary examination of the existing product, the computer-aided design can be applied.

8.3.2 Phase change material selection

Before appropriate phase change materials can be selected and applied in a necessary quantity, the heat flux from the body through the ski wear ensemble into the environment under different ambient temperatures needs to be determined.

Ski wear ensembles are normally worn in the wintertime in the mountains under low ambient temperatures. Thus, for such an application of the ski wear ensemble, an ambient temperature range between -15°C and + 5°C has been considered.

Furthermore, a mean skin temperature of 33°C has been used in the calculations. A mean skin temperature between 31 °C and 34 °C represents a comfort sensation for the human body. When the skin temperature exceeds 34°C, the body perspires through the skin glands to cool the skin down. On the other hand, a drop in skin temperature below 31°C results in a cold sensation which leads to shivering.

The temperature profiles are calculated based on a steady-state (dry and wet) heat transfer through the garment layers. Air layers of one to two millimeters between the garment layers were considered when calculating the temperature profiles. The temperature profiles for the given ambient temperatures are shown in Fig. 8-3.

The temperature profiles shown in Fig.8-3 determine the application temperatures inside the garment layers. The selected phase change materials need to operate in these temperature ranges to obtain the desired thermal effects.

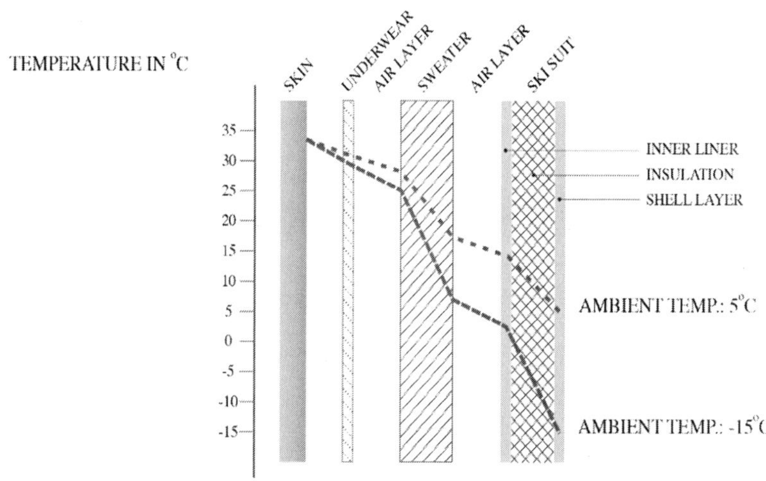

Fig. 8-3: Temperature profiles through the ski wear ensemble.

The temperature ranges in which a phase change material should absorb and release latent heat when applied to these ski wear components are summarized in Table 8-3.

Ski wear ensemble	Temperature range for the latent heat absorption of the phase change material, °C	Temperature range for the latent heat release of the phase change material, °C
Thermal underwear	33-37	30-32
Fleece sweater	26-31	21-25
Ski suit liner	16-20	8-14

Table 8-3: Appropriate temperature ranges for the absorption and release of latent heat by the phase change material located in the different components creating the ski wear ensemble.

If applied to the thermal underwear, the phase change material's purpose is mainly to absorb surplus heat during strenuous activities. Thus, the phase change material selection for this application should focus on latent heat absorption of the phase change material taking place mainly in a temperature range between 33 °C and 37°C.

Applied to the fleece sweater, the phase change material should provide an active thermal barrier effect which consists of absorbing latent heat during strenuous activities and releasing latent heat during rest periods under low ambient temperatures. Latent heat should be absorbed by the phase change material when the temperature in this location rises above 26°C. The stored latent heat should be released by the phase change material when the temperature drops below 25 °C.

The active thermal barrier function generated by the phase change material application in the ski suit (jacket) between the inner liner and the insulation layer is established when latent heat is absorbed in a temperature range between 16 °C and 20 °C and is released in a temperature range between 8°C and 14 °C.

Phase change material should not be located between the insulation layer and the shell layer because it would be too far away from the wearer's body and, therefore, not efficient for keeping the skin/body temperature in a thermal comfort range.

Based on the temperature ranges summarized in Table 8-3, phase change materials are selected which operate in these temperature ranges. The phase change material selection is supported by a database that comprises about 200 materials, such as paraffins, fatty acids, salt hydrates, eutectics, etc., from various suppliers. Microencapsulated paraffins are also contained therein. The database includes operation temperature ranges, latent heat storage capacities, and information regarding properties such as thermal stability, toxicity as well as prices and availabilities. The database is part of the computer-aided calculation and simulation procedure and is continuously updated.

By utilizing the database, the following microencapsulated paraffins are pre-selected for an application in the ski suit ensemble:

- Eicosane microcapsules in the thermal underwear.
- Octadecane microcapsules in the sweater.
- Hexadecane microcapsules between the inner liner and the insulation batting of the ski suit jacket.

8.3.3 Estimation of the phase change material quantity

The heat input into the product must be considered to determine the sufficient latent heat storage capacity and, hence, the necessary phase change material quantity. In garment applications, the heat input is mainly related to the amount of heat generated by the wearer's body. To avoid an increase in the body's core temperature, the heat emitted from the body must be released entirely through the garment layers into the environment.

The heat produced by the human body depends on the wearer's activities. The metabolic heat rates that characterize the heat generation of the human body during certain activities have been determined using an oxygen consumption method in field tests carried out over several years.

Another database, which is also part of the computer-aided calculation and simulation procedure, contains about 150 metabolic heat rates. The database summarizes metabolic heat rates for basic activities such as sitting, standing, walking, and running at different speeds and for activities related to different sports or work duties.

Skiing comprises very different activities for which metabolic heat rates have been determined. Some of them are summarized in Table 8-4. Because of variations in the activity levels, ranges are given.

Activity	Metabolic heat rates, kJ/min.m^2
Standing	4-5
Sitting	3-4
Walking and carrying the ski	6-10
Downhill skiing	9-17
Cross-country skiing	10-20

Table 8-4: Metabolic heat rates of the human body during skiing activities.

The high heat generation during the downhill and cross-country skiing activities could easily lead to an overheating of the body. In contrast, the metabolic heat rates generated by the body when waiting for the ski lift over a longer period or sitting on the lift are comparatively low and could result in hypothermia.

To estimate the necessary phase change material quantity, computer simulations are carried out. An excerpt of one hour is taken from a five-hour study to demonstrate the procedure. The following activities have been considered in this one-hour part of the simulation:

1: Walking from the parking lot to the ski lift in 10 minutes at an ambient temperature of -10 °C.
2: Standing in line and waiting on the ski lift for 12 minutes at an ambient temperature of -10 °C.
3: Sitting on the ski lift for 10 minutes at an ambient temp. of -10 °C.
4: Downhill (DH) skiing for 10 minutes at an ambient temperature of - 5 °C.
5: Standing for 6 minutes at an ambient temperature of -7 °C.
6: Cross-country (CC) skiing for 12 minutes at an ambient temperature of -7 °C.

Heat balance calculations have been carried out to determine the necessary amount of latent heat absorption or latent heat release by the phase change material integrated into the ski wear ensemble, The results of these heat balance calculations received in the simulation of these skiing activities are summarized in Table 8-5.

No.	Ambient Temp, °C	Activity	Duration, min.	Heat generation in the upper body area, kJ	Heat transfer through the ski jacket, kJ	Heat absorption (+)/ heat release (-) of the phase change material kJ
1	-10	Walking	10	70	61	+9
2	-10	Standing	12	48	73	-25
3	-10	Sitting	10	30	61	-31
4	-5	DH skiing	10	110	53	+ 57
5	-7	Standing	6	30	34	-4
6	-7	CC skiing	12	132	67	+65

Table 8-5: Results of the heat balance calculations for a skiing activity simulation.

Columns 2, 3, and 4 of Table 8-5 comprise the prevailing ambient temperatures during the different steps, the activities being simulated in these steps, and their durations.

Column 5 of Table 8-5 contains the amounts of heat generated by the human body in the upper chest area related to the specific activities. Metabolic heat rates like those shown in Table 8-4 were considered. The upper chest area is taken into consideration because the phase change material has been intended to apply only to the garment components worn in the upper part of the body such as the thermal underwear, the sweater, and the ski jacket.

The amounts of heat, which is then removed by the dry and wet heat transfer through the ski suit layers in the upper part of the body based on the temperature gradient between the skin temperature of 33 °C and the prevailing ambient temperature, are summarized in column 6 of Table 8-5. To calculate the numbers included in column 6, the test results of the preliminary tests of the ski suit components summarized in Table 8-2, as well as the temperature profiles summarized in Fig. 8-3 are utilized.

Finally, column 7 of Table 8-5 contains the amounts of heat that the phase change material needs to handle to avoid temperature spikes in the microclimate. The plus sign in column 7 indicates the necessity for the phase change material to absorb latent heat, while the minus sign expresses that the phase change material needs to release part of its stored latent heat.

The computer simulations are carried out several times with variations in the single activities, their durations, and the ambient temperature conditions.

Based on the results of these simulations, a latent heat storage capacity of about 12 kJ has been contrived to be sufficient for temperature regulation in these garment components worn in the upper part of the body.

8.3.4 Selection of a suitable textile substrate

The selection of a suitable textile substrate used as a carrier of the phase change material is supported by another database of available products integrated into the calculation and simulation procedure. This database includes the material composition of the substrates and the way the substrates are made. Furthermore, the database summarizes specific parameters of the substrates, such as weight per unit area, thickness, density, thermal resistance and evaporative resistance, water vapor absorption, and mechanical properties, all of which are necessary for successful product development. The database is continuously updated.

To select a suitable substrate for the phase change material application in the ski suit ensemble, the materials used in the components and the construction of the components need to be considered.

The thermal underwear is made of a knit of wool. Currently, wool or even cotton fibers equipped with phase change material do not exist. A coating of one side of the thermal underwear fabric would reduce the knit's moisture management substantially. Therefore, the thermal underwear does not seem to be a suitable substrate for the phase change material application in the ski wear ensemble.

The sweater is made of a polyester fleece. Polyester fibers equipped with phase change materials are commercially available. However, the required latent heat storage capacity for the ski wear ensemble, could not be realized in such a sweater. Therefore, the sweater is not a suitable substrate for the phase change material application as well.

The ski suit itself consists of a thin polyamide 6.6 (nylon) fabric as the inner liner, a polyester batting as the insulation layer, and a polyamide shell fabric. The best location for the phase change material application is between the inner liner and the insulation layer. The dense structure of the ski suit's inner liner ensures a fast heat transfer to the phase change material and, therefore, an effective absorption of excessive heat. Because of the heat flux direction from the body into the environment, the latent heat stored in the phase change material is released into the insulation layer. The lofty batting used in the insulation layer allows for an extended stay of the latent heat released by the phase change material which supports the desired thermal barrier function sufficiently and prevents heat loss from the body during non-strenuous activities on the ski slopes (Configuration 2 in 7.6).

In the design, a lightweight and thin nonwoven scrim screen-printed with a compound containing hexadecane microcapsules could, for example, be inserted between the inner liner and the insulation layer of the ski suit. Based on the latent heat storage capacity of the hexadecane microcapsules of about 190 J/g, the application of a microcapsule quantity of 65 grams would be sufficient.

8.3.5 Design of the end-use product

Apart from devising and realizing an attractive design, the optimization of the phase change material's thermal effects is the goal of this last step. The computer-aided calculation and simulation procedure includes various tools to support the final design process.

Studies have shown that it is often not necessary to put phase change material in all parts of the end-use product. Applying phase change material to the areas that are problem zones from a thermal standpoint and regulating the heat flux through these areas is often enough.

In some end-use products, it is an advantage to use phase change materials operating in different temperature ranges in various locations of the product. The heat flux from one location where one distinct phase change material is arranged to another location where a different phase change material is placed often leads to an extended duration of the desired thermal effects. Different amounts of heat generated in different parts of the human body often require the application of different phase change material quantities in selective garment locations.

To identify the thermal problem areas, heat flux measurements, and infrared photographs are taken while the original end-use product is investigated. These tools are integrated into the program and extensively used in the design process.

The infrared photographs and the heat flux measurements indicate that the highest amount of heat is built up in the upper chest, neck, and head area during strenuous activities. The skin temperature in this part of the body increases substantially even under low ambient temperature. On the other hand, because of the low blood circulation in the extremities, the temperature drop in these areas of the body is very fast under low ambient temperatures.

The phase change material should be mainly located in the lower chest area and around the arms to keep the ski jacket wearer warm. Fig. 8-4 shows the most suitable arrangement of phase change material in a ski jacket.

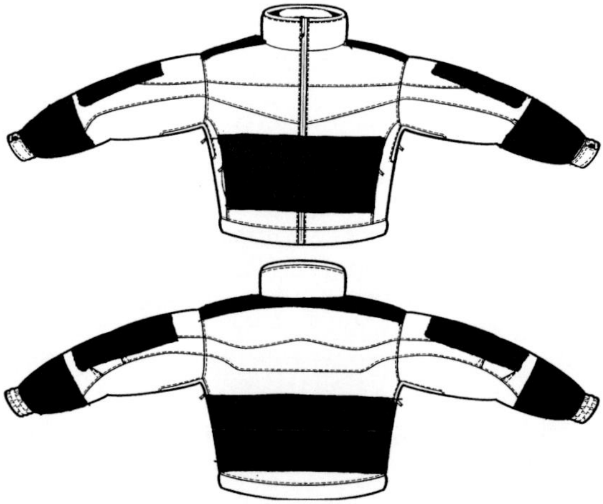

Fig. 8-4: Suitable phase change material arrangements in a ski jacket in the front and in the back (marked in black).

References

1. Pause, Barbara. 2001. "Computer-optimized design of thermo-regulated activewear". Technical Textiles International. No. 11: 23-26
2. Pause, Barbara. 2017. "Smart Textiles Created by the Application of Phase Change Material (PCM)". ITMC International Conference on Intelligent Textiles and Mass Customization, Ghent, Belgium, October 15-17, 2017
3. Pause, Barbara. 2010. "Phase change materials and their application in coatings and laminates for textiles". In: Smart textile coatings and laminates, edited by Smith, W.C., 236-250. Woodhead Publishing Ltd.

4. American Society for Testing and Materials (ASTM) International. 2020. "Standard Test Methods for Mass per Unit Area (Weight) of Fabric". Standard D3776/D3776M-20. Book of Standards, Vol. 07.01
5. American Society for Testing and Materials (ASTM) International. 2019. "Standard Test Method for Thickness of Textile Materials". Standard D1777-96. Book of Standards, Vol. 07.01
6. American Society for Testing and Materials (ASTM) International. 2023. "Standard Test Method for Thermal Resistance, Evaporative Resistance, and Total Heat Loss Measurements of Clothing Materials Using a Sweating Hot Plate". Standard F1868-23. Book of Standards, Vol. 11.03

CHAPTER NINE

APPLICATION OF PHASE CHANGE MATERIAL FOR THERMAL PROTECTION AND THERMO-PHYSIOLOGICAL COMFORT IMPROVEMENT

9.1 Principles

Phase change materials are suitable for protection against external heat exposure. Under these circumstances, the phase change material will absorb excessive heat and will limit and delay the heat transfer into the interior. Examples of such applications are various protective means against heat exposure.

The application of phase change material can also limit the transfer of heat generated by the body into the environment. Thus, the body can be protected against heat loss in a cold environment, such as a cold store.

In addition to its protective function, the phase change material is often used to enhance the thermo-physiological comfort of an existing product. Applications where the phase change material provides a thermo-physiological comfort improvement can be found in sports equipment, medical supplies, or bedding products. In these applications, the temperatures surrounding the phase change material are kept as constant as possible in a comfortable range by taking advantage of its thermoregulating feature. When the skin temperature of the human body remains constant in the comfort zone, perspiration is limited which further enhances the comfort sensation.

In this chapter, the phase change material's protective function and contribution to thermo-physiological comfort will be discussed utilizing the examples of heat resistant gloves, cold protective garments, and mattresses.

9.2 Heat-resistant gloves

Wearing conditions

Heat-resistant gloves are used, for instance, by firefighters, welders, or steel mill workers, to protect their hands against burn injuries when exposed to contact heat or radiant heat. The duration of the heat exposure varies widely, and often a heavy and hot object needs to be carried from one location to another. Fig. 9-1 shows a steel mill worker lifting and carrying a hot disk. While in contact with the hot disk, the heat emitted from the disk enters the glove's surface and penetrates through the glove. Due to the heat penetration, the temperature inside the glove and, hence, the skin temperature rises. If the skin temperature increases to 45 °C, the wearer experiences a pain sensation. A further rise in the skin temperature to 55°C and above leads to burn injuries.

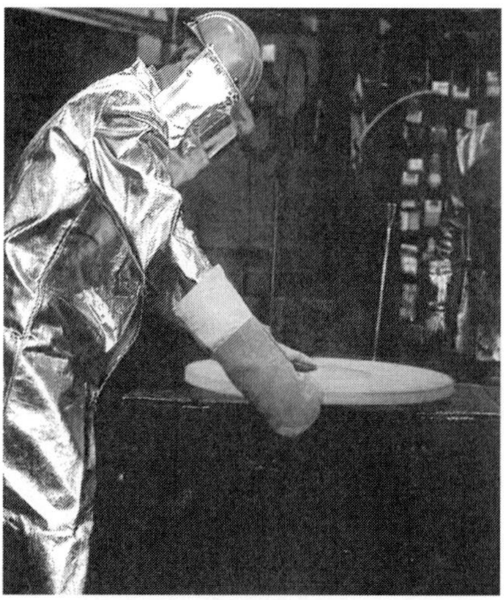

Fig. 9-1: Contact with a hot object by a steel mill worker.

To protect against such health hazards, the glove must provide significant insulation, often requiring the use of a relatively thick insulation material. However, this approach makes the glove very bulky. Hand dexterity is likely reduced when a glove with such a thick, bulky insulation layer is used.

Objectives for the phase change material application in heat-resistant gloves

Applying phase change material in heat-resistant gloves shall increase thermal protection against contact and radiant heat by absorbing excess heat starting at a given trigger temperature. The trigger temperature and the location of the phase change material in the glove configuration should be chosen so that a rise in the glove's interior temperature above 45 °C is prevented over an extended period. In addition, an attempt should be made to reduce the thickness of the insulation layer without compromising the thermal protection effect. The decrease in thickness will make the glove less bulky enhancing the hand dexterity [1].

Heat-resistant glove design

An example of a heat-resistant glove used as the basic configuration in the development is shown in Fig. 9-2. The palm area that comes into contact with the hot object usually consists of the following fabric components:

- Aramid shell.
- Fiberglass mat insulation.
- Fire-resistant cotton fabric as a liner.

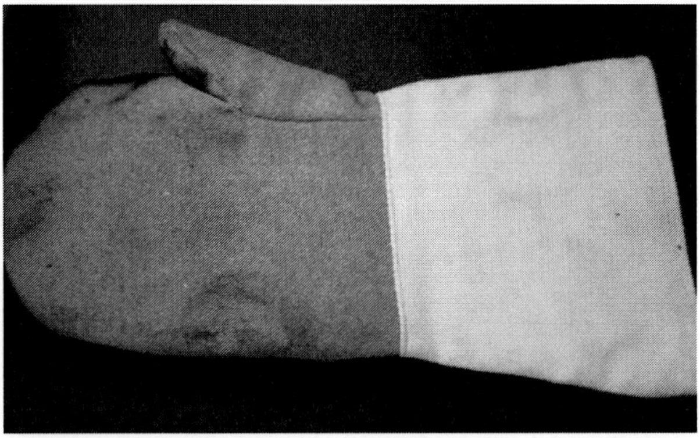

Fig. 9-2: Heat-resistant glove.

Phase change material application

A suitable location for the phase change material application in the heat-resistant glove is the backside of the aramid shell fabric because the incoming heat flux is immediately blocked due to the absorption of latent heat which prevents the temperature rise for a certain period. When the phase change material is completely charged, the further heat flux into the glove is limited by the insulation. Preferably, a non-combustible phase change material operating in a temperature range slightly above 100 °C is embedded into a heat-resistant coating compound such as a silicone rubber coating compound or a ceramic coating compound, which is topically applied to one side of the aramid fabric. A picture of an aramid fabric coated on one side is shown in Fig. 9-3.

The reduction in the breathability of the fabric due to the coating is not a problem because the gloves are mostly only worn for a short period of time.

Fig. 9-3: Uncoated aramid fabric (left) and aramid fabric with a silicone rubber coating containing phase change material (right).

The application of 140 g/m^2 of a phase change material with a latent heat storage capacity of 285 J/g leads to a total latent heat storage capacity of the coated aramid fabric of about 40 kJ/m^2. The coating enhances the thickness of the aramid fabric by one millimeter. Despite this increase in the thickness of the shell layer, the overall thickness of the glove was reduced by 1.8 millimeters by replacing the thick insulation with a thinner variant. The results of the thermal insulation measurements carried out on the original glove design without phase change material and the new glove design with phase change material by utilizing the test method described in Chapter 6.3 are summarized in Table 9-1.

Thermal insulation data	Original glove design	New glove design
Overall thickness, mm	7.8	6.0
Basic thermal resistance, m^2K/W	0.0804	0.0627
Dynamic thermal resistance, m^2K/W	-	0.0773
Total thermal resistance, m^2K/W	0.0804	0.1400

Table 9-1: Thermal insulation data of the glove configurations.

The results summarized in Table 9-1 indicate that the basic thermal resistance decreased by 22 percent when the glove's thickness was reduced from 7.8 millimeters to 6 millimeters. However, the total thermal resistance (insulation) has increased by more than 40 percent due to dynamic thermal resistance contributed by the latent heat absorption of the phase change material, the new glove design is equipped with.

The insulation effect obtained by utilizing the heat-resistant glove normally decreases when the glove's insulation layer is compressed while carrying a heavy, hot object. This is another benefit of the thermal insulation established by the phase change material. The dynamic thermal resistance provided by the phase change material is not compromised when the layers the glove consists of are compressed. A comparison of the thermal resistance values of both glove designs with a compression (2.0 kPa) and without compression (0.2 kPa) is presented in Fig. 9-4.

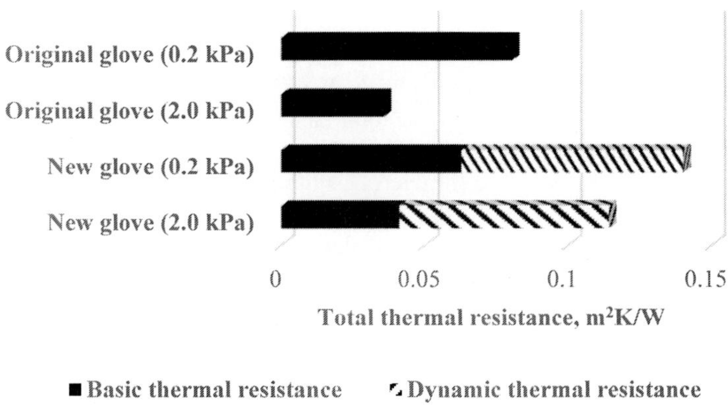

Fig. 9-4: Comparison of the thermal resistance of both glove variants with and without compression.

Test of the thermal protection against contact heat

Both glove configurations were tested regarding their protection against contact heat exposure. The tests were carried out utilizing the ISO 12127 test method [2]. In this test, the test sample is brought into contact with the hot surface of a heated cylinder, and the temperature increase on the back side of the sample is measured over time. To quantify the degree of protection, a threshold time t_T is determined. The threshold time t_T represents the time in which the temperature on the back side of the sample has increased by 10 K after contact with the hot surface. In the test, the outer aramid shell encounters the hot surface, and the temperature measurements are taken on the liner's backside as the glove's inside. The tests were carried out at contact temperatures of 100 °C, 200 °C, 300 °C, 400 °C, and 500 °C. Furthermore, the tests were performed under contact pressures (CP) of 0.2kPa and 2 kPa [3]. The threshold times t_T are given in seconds. The results are summarized in Table 9-2.

Test material	CP, kPa	100 °C t_T,, s	200 °C t_T,, s	300 °C t_T, s	400 °C t_T, s	500 °C t_T, s
Original glove	0.2	82	58	40	24	11
Original glove	2.0	53	37	23	14	8
New glove	0.2	127	105	86	57	26
New glove	2.0	112	88	67	44	21

Table 9-2: Thermal protection against contact heat.

The test results summarized in Table 9-2 indicate that the new glove design equipped with phase change material provides significantly higher protection against contact heat than the original glove design with and without the applied contact pressure (compression). The increase in protection against contact heat due to the phase change material application is about 2 to 3 times. The heat protection is improved despite the reduced thickness of the new glove design. This improvement is caused by the thermal barrier effect due to the latent heat absorption of the phase change material.

9.3 Cold protective garments

Wearing conditions

Cold protective garments are used to protect those working under extremely low ambient temperatures against cold stress, injuries, and hypothermia. It has been estimated that in Germany alone, about one million people work under such conditions.

Most of these workplaces are in cold stores where temperatures range from –18 °C to –30 °C. Workers in the cold stores normally carry out three major tasks: sorting and distribution work, transportation work, and inspections. The proportions in which the different tasks contribute to the total workload are the following: 54 percent sorting and distribution work, 40 percent transport work, and 6 percent inspection [4].

Collecting the appropriate goods, taking these packages from the storage palettes, and manually transporting them to the forwarding palettes are all tasks assigned to the sorting and distribution workers. Holding and carrying packages weighing up to 15 kilograms requires a high degree of physical exertion, leading to heat buildup in the microclimate and an increase in the skin temperature. The rising skin temperature initiates sweat production of the skin glands, which can lead to a thorough soaking of the garments being worn, finally resulting in a decrease in the thermal insulation capacity of the garment components. The entire sorting and distribution work is done in a stable climate at very low temperatures below -18°C. Fig. 9-5 shows such distribution work, carried out in a cold store.

Fig. 9-5: Sorting and distribution work in a cold store.

After an exposure time of about 90 minutes, the worker needs to interrupt his work in the cold for about 30 minutes to avoid hypothermia. During this break, the cold protective garment is normally dried and warmed up in an oven.

The transport of goods from the loading ramp into the cold store and vice versa is done with forklifts. Operating the forklifts requires a minimum of physical exertion which causes the operator's body to generate a relatively small amount of heat. If this work had to be done only in the cold store, the worker's skin temperature would drop substantially already in a short period, and he would experience a cold sensation very quickly. But driving a forklift in a cold store is carried out while changing steadily between areas of different ambient temperatures - the loading ramp and the cold store. The steady temperature change results in a significant stress situation the workers are suffering from. This work is also normally interrupted after 90 minutes for a 30-minute break to warm up again [5]. Fig. 9-6 shows the operation of a forklift in the cold store.

Fig. 9-6: Operation of a forklift in a cold store.

Objectives for the phase change material application in cold protective garments

The temperature changes during the forklift driving and the different physical exertions during the sorting and distribution work make a phase change material application in cold protective garments especially worth considering. In other words, without using the active thermal barrier function of phase change material in a cold protective garment, the thermal insulation effect appropriate for the different work conditions can only be obtained by adding or removing garment layers very frequently which is impossible.

In a cold protective garment application, the phase change material provides an active thermal barrier effect which acts in addition to the passive thermal (insulation) barrier of the garment system. Thus, the phase change material will absorb surplus heat generated by the body during the sorting and distribution work. The perspiration which results from a rise in the skin temperature during such strenuous work will then be reduced. The absorbed heat will be released, for instance, when driving a forklift or doing some inspection work where the physical activity is limited. The phase change material application will increase thermal protection against cold and improve thermo-physiological wearing comfort. As a result of the increase in thermal protection, hypothermia will be avoided.

Finally, the improved thermal protection will lead to a possible extension in the working periods without the need for an intermediate warm-up [6]. This would increase productivity in the working process.

Cold protective garment design

Cold protective garments (mainly overalls) consist of three layers: an inner liner, an insulation layer, and a shell layer. Under their cold protective garments, the cold store workers wear thermal underwear and a normal work suit [7]. The cold protective garment ensemble, which was used for the phase change material selection, consists of the following components:

- Thermal underwear (100 percent wool).
- Work suit (65 percent cotton, 35 percent polyester).
- Cold protective suit:
  - Inner liner (100 percent polyamide 6.6),
  - Insulation batting (100 percent polyester),
  - Shell layer (100 percent polyamide 6.6).

In the cold protective suit, the insulation batting is quilted between the liner fabric and the shell fabric.

The weight and thickness of the cold protective garment components are included in Table 9-3. They were determined in accordance with ASTM Standard D3776/D3776M and the ASTM Standard D1776-96 [8, 9]

Cold protective garment components	Weight per unit area, g/m^2	Thickness, mm
Thermal underwear	195	1.3
Work suit	290	0.7
Inner liner	70	0.2
Insulation batting	210	16.0
Shell layer	200	0.3

Table 9-3: Weight and thickness of the cold protective garment components.

The dry and wet heat transfer through these garment layers is determined by measuring their thermal resistance and evaporative resistance values following ASTM Standard F1868-23 [10].

The test results are summarized in Table 9-4.

Cold protective garment components	Thermal resistance, m^2 K/W	Evaporative resistance, m^2 Pa/W
Thermal underwear	0.0228	3.5
Work suit	0.0049	3.6
Inner liner	0.0009	3.8
Insulation batting	0.5674	3.3
Shell layer	0.0017	4.9

Table 9-4: Thermal resistance and evaporative resistance values of the cold protective garment components.

Phase change material application

The calculation of the temperature profiles through the cold protective garment systems utilizing the computer-aided calculation and simulation procedure described in chapter 8.3.2 has been carried out for an ambient temperature range anticipated for cold stores between –18 °C and - 30°C. Because of the goal to keep the skin temperature in the comfort range, the calculations were made considering a mean skin temperature of 33 °C. The temperature profiles have been calculated based on a steady-state (dry and wet) heat transfer through the garment layers. Air layers of one to two millimeters between the garment layers were considered when calculating the temperature profiles. The temperature profiles are shown in Fig. 9-7.

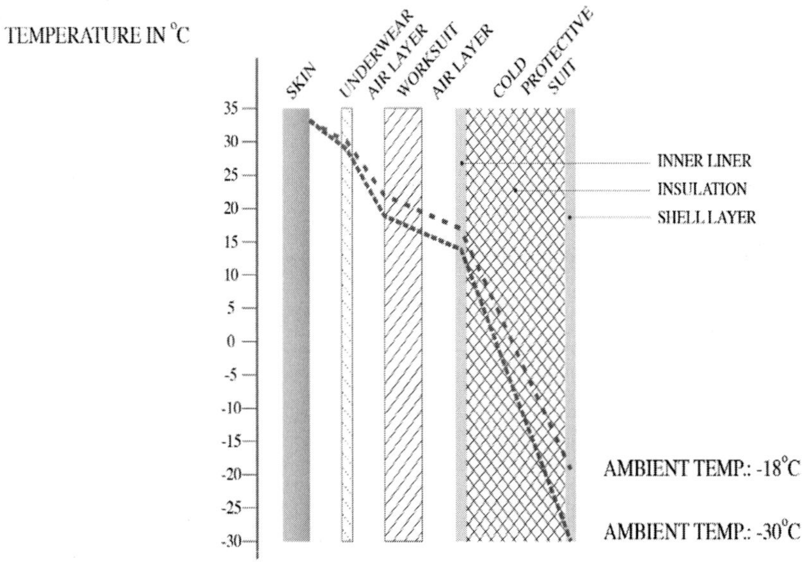

Fig. 9-7: Temperature profiles through the cold protective garment ensemble.

To enhance thermal protection and obtain sufficient thermoregulating effects, the phase change material should be placed inside the cold protective garment between the inner liner and the insulation batting. In this arrangement, a phase change material needs to be selected which absorbs latent heat when the temperature at this location rises above 17 °C and releases latent heat when the temperature drops below 12°C. A microcapsule containing a hexadecane paraffin was selected from the database connected to the computer-aided calculation and simulation procedure which would be suitable for this application. The commercially available hexadecane microcapsule possesses a latent heat storage capacity of 190 J/g.

To calculate the necessary amount of the microencapsulated phase change material applied to the cold protective garment system, the computer-aided calculation and simulation procedure described in Chapter 8.3.3 was applied. In this procedure, activity simulations are performed in which the heat generation of the worker's body while carrying out specific tasks and the heat release through the layers of the cold protective garment ensemble are considered. Based on this information, the necessary latent heat absorption/latent heat release by the microencapsulated phase change material applied to the cold protective garment is determined.

The heat generated by the worker's body while performing the different tasks in the cold stores is given in metabolic heat rates. These metabolic heat rates are summarized in Table 9-5. Because of variations in the activity levels, ranges are given.

Tasks	Metabolic heat rates, kJ/min.
Walking	7-9
Inspection	4-5
Forklift driving	4-6
Sorting and distribution work	10-12

Table 9-5: Metabolic heat rates for tasks carried out in cold stores.

The simulations considered a working period of 90 minutes. The warming period following the working period was not part of the simulation. Computer simulations are usually carried out multiple times, and the working tasks and their durations vary in these simulations.

One of the simulations covered the following work tasks:

1: Walking from the warming locker room to the storage area for 5 minutes at a temperature of -18 °C.
2: Inspection of the goods in the storage area for 3 minutes at a temperature of -18 °C.
3: Sorting and distribution work in the storage area for 25 minutes at a temperature of -18 °C.
4: Forklift driving (FL) for 20 minutes at a temperature of -18 °C.
5. Inspection of the goods in the freezer room for 5 minutes at a temperature of -30 °C.
6: Sorting and distribution work in the freezer room for 12 minutes at a temperature of -30 °C.
7: Forklift (FL) driving in the storage area for 15 minutes at a temperature of -18 °C.

8: Walking from the storage area to the warming locker room for 5 minutes at a temperature of -18 °C.

Based on the activity simulations, heat balance calculations have been carried out. The results of these calculations are summarized in Table 9-6.

No	Ambient temp., °C	Activity	Duration, min.	Heat generation of the entire body, kJ	Heat transfer through the garment, kJ	Heat absorption (+)/ heat release (-) of the phase change material, kJ
1	-18	walking	5	40	39	+1
2	-18	inspection	3	12	23	-11
3	-18	sort./dist.	25	275	193	+82
4	-18	fl driving	20	100	154	-54
5	-30	inspection	5	20	48	-28
6	-30	sort./dist.	12	132	114	+18
7	-18	fl driving	15	75	116	-41
8	-18	walking	5	40	39	+1

Table 9-6: Results of the heat balance calculations for working in the cold store.

Columns 2, 3, and 4 of Table 9-6 comprise the temperatures in the different locations of the cold store, the work tasks carried out in these locations, and their durations.

Column 5 of Table 9-6 contains the amounts of heat generated by the human body performing these specific tasks. Metabolic heat rates summarized in Table 9-5 were considered.

The amounts of heat that are then removed by the dry and wet heat transfer through the cold protective garment layers based on the temperature gradient between the skin temperature of 33 °C and the prevailing ambient temperature in the different locations are summarized in column 6 of Table 9-6. To calculate the numbers included in column 6, the test results summarized in Table 9-4 and the temperature profiles summarized in Fig. 9-7 are utilized.

Finally, column 7 of Table 9-6 contains the amounts of heat that the phase change material needs to handle to avoid temperature spikes in the microclimate. The plus sign in column 7 indicates the necessity for the phase change material to absorb latent heat, while the minus sign expresses that the phase change material needs to release part of its stored latent heat.

The heat balance calculations indicate that a latent heat storage capacity of 16 kJ/m^2 is required to keep thermal comfort while working under such harsh climatic conditions and avoid a drop in the wearer's skin temperature, which could lead to hypothermia during the 90-minute work in the cold store. Under consideration of the application of microencapsulated hexadecane possessing a latent heat storage capacity of about 190 J/g, the utilization of about 85 g/m^2 of the microencapsulated hexadecane is necessary. Covering the entire cold protective suit, a latent heat storage capacity of about 29 kJ needs to be generated, requiring the use of a quantity of about 150 g of microencapsulated hexadecane. The microcapsules are integrated into a binder and applied to a nonwoven scrim in a screen-printing procedure. The scrim is then inserted into the cold protective suit and quilted between the inner liner fabric and the insulation batting.

In simulations where the working period was extended to 120 minutes, it was calculated that a latent heat storage capacity of 20 kJ/m^2 was required to maintain thermal comfort and avoid hypothermia. This leads to an increase in the necessary phase change material quantity to 105 g/m^2 applied to the cold protective suit. Extending the working period from 90 minutes to 120 minutes without interruption would result in increased productivity.

Test of the cold protective garment in a cold store

Cold protective garments with and without phase change material were tested in a cold store. The following tasks were performed in a 60-minute test:

1. 20 minutes forklift driving inside the cold storage area at - 18°C.
2. 20 minutes lifting and carrying packages weighing 15 kg inside the cold storage area at - 18°C.
3. 20 minutes forklift driving between the loading deck at -5°C and the cold storage area at - 18°C.

The microclimate temperature between skin and thermal underwear was measured in several locations during this test, and the results were averaged. The test results are summarized in Fig. 9-8.

As indicated in Fig. 9-8, applying phase change material to the cold protective garment prevents the microclimate temperature from dropping below uncomfortable 30 °C, avoiding large temperature swings.

In contrast, while wearing the cold protective garment without phase change material, the microclimate temperature drops dangerously below 28 °C in the first part of the test. The low activity while driving the forklift at a very low temperature of – 18 ºC causes this temperature decrease.

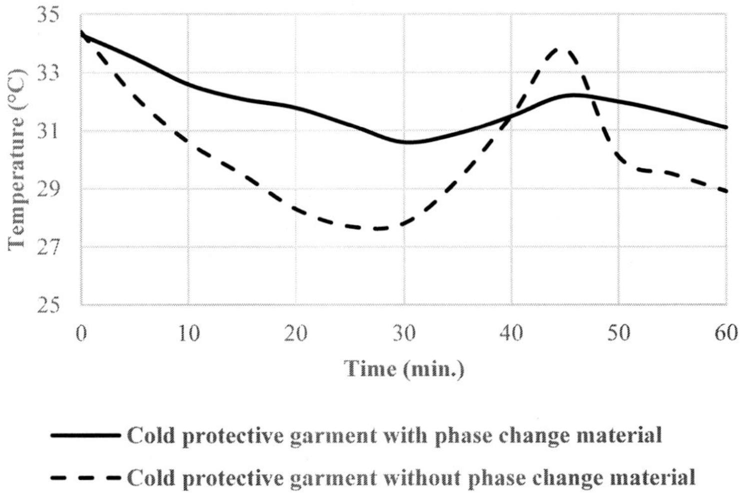

Fig. 9-8: Test of cold protective garment ensembles with and without phase change material application in a cold store.

The temperature in the microclimate rises to 34 ºC when the worker lifts and carries heavy packages. However, this fast rise in skin temperature also leads to a substantial amount of perspiration. This perspiration penetrates the thermal underwear and the work suit worn underneath the cold protective garment and will reduce their thermal insulation properties. This drop in insulation is mainly responsible for the significant decrease in the microclimate temperature at the end of the test.

9.4 Bedding products

Conditions of use

Thermo-physiological comfort during sleep is one of the critical factors affecting sleep quality besides spine alignment, contact pressure, weight distribution, and water vapor exchange. Thermo-physiological comfort is obtained when the skin temperature remains between 31 °C and 34 °C during sleep. Keeping the skin temperature in the comfort range limits the amount of perspiration generated by the skin glands [11].

Bedding products such as mattresses, mattress covers, sheets, cushions, or comforters are used under room-temperature conditions, i.e., temperatures between 15 °C and 25 °C. The room temperature can change during sleep if ambient air penetrates through the windows into the bedroom. The bedding components are at room temperature when not in use.

Objectives for the phase change material application in bedding products

The phase change material application in bedding products should enhance thermo-physiological comfort. In most circumstances, this thermo-physiological comfort is related to a cooling sensation [12].

The human body generates a metabolic heat rate of about 85 Watts while sleeping [13]. 90 percent of this heat is emitted through the skin and must be transferred into the environment to avoid overheating of the skin and the body. Because many of the bedding products possess high thermal insulation, the heat transfer through them is very limited [14]. Mattresses are the bedding products that provide the highest insulation against the heat transfer away from the body. Therefore, they should be considered first and foremost for a phase change material application. A phase change material applied to the upper part of the mattress provides a cooling sensation by absorbing such excessive heat as soon as the temperature in the microclimate between the skin and the mattress reaches a certain trigger point. The latent heat absorption of the phase change material keeps the microclimate temperature constant in the comfort range and limits the perspiration by the skin glands [15, 16].

Mattress design

Most mattresses nowadays consist of three layers – a support core layer, a transitional layer, and a comfort layer. The support core at the bottom of the mattress is intended to stabilize the entire mattress. Support core layers consist of either steel coils, air chambers, dense polyfoam, or latex. The transitional layer lies on top of the support core and below the comfort layer. The purpose of this layer is to reinforce the comfort layer and to provide a buffer between the base material and the human body. The transitional layer is made of polyfoam or latex. As the uppermost layer of the mattress construction, the comfort layer is either made of memory polyurethane foam, polyfoam, or latex.

The comfort layer can also be a mattress topper for an existing mattress. The study used a latex mattress topper made of one hundred percent natural latex. The mattress topper is 1.9 meters by 1 meter and 0.025 meters thick and weighs 3.7 kg.

Phase change material application

To provide a sufficient cooling sensation, the phase change material should be placed not far away from the skin. Therefore, a suitable phase change material application would be on top of the mattress topper. The mattress topper would then be covered by the bed sheet.

A phase change material has been selected which absorbs latent heat in a temperature range between 28 °C and 33 °C. This temperature range ensures that the microclimate temperature remains in the comfort range below 34°C. The selected phase change material releases latent heat in a temperature range between 22 °C and 26 °C. This temperature range of the latent heat release makes a sufficient recharge of the phase change material possible under circumstances where the mattress is not covered by the body.

Paraffin octadecane operates in this temperature range. The latent heat storage capacity of microencapsulated octadecane is about 190 J/g. The phase change material was mixed into a coating compound in a 50 wt. % quantity and topically applied to one side of the mattress topper by a spray-coating procedure.

200 g/m^2 coating compound was applied to the mattress topper increasing its weight by about 10 percent. The 50 wt.% phase change material application in the coating compound results in a latent heat storage capacity of about 19 kJ/m^2. The phase change material spread over the entire surface of the mattress topper provides a latent heat storage capacity of about 36 kJ.

Test of the thermo-physiological comfort

To quantify the thermo-physiological comfort improvement, mattresses equipped with the topper with and without phase change material treatment were tested against each other. The tests were performed with a heated and sweating torso form in a climatic chamber test setup. The temperature inside the climatic chamber was 18 °C during the tests.

The heat emission from the torso form's surface was realized by simulating the heat emission of the human body while lying at rest. In addition, the torso form releases water vapor through tiny pores on its surface. The amount of water vapor released during the test varies depending on the torso's surface temperature.

During the test, the torso form was laid on top of a mattress topper that was treated or untreated with phase change material and covered with a bed sheet.

Temperature/humidity sensors measured the temperature development and the moisture generation in the microclimate between the bed sheet and the torso form. The test results received for the temperature development in the microclimate within one hour are shown in Fig. 9-9.

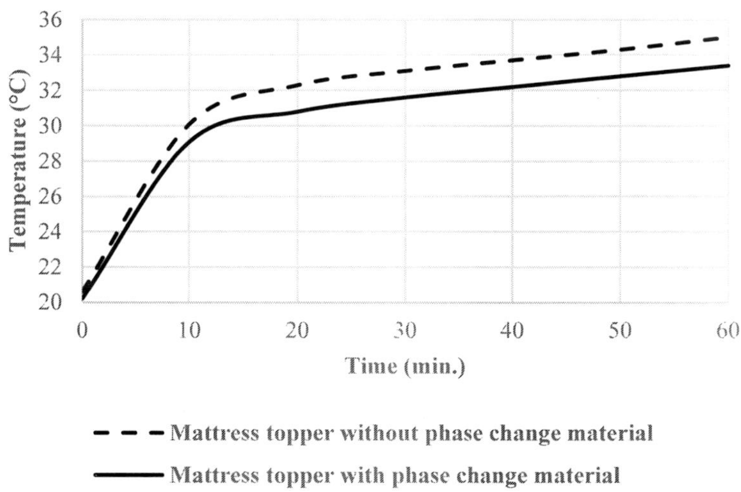

Fig. 9-9: Temperature development in the microclimate.

The results summarized in Fig. 9-9 indicate that the microclimate temperature above the topper equipped with phase change material stays entirely in the comfort range below 34 °C during the one-hour test. In contrast, the microclimate temperature measured for the topper without phase change material treatment exceeds a comfortable temperature of 34°C already after about 40 minutes into the test.

The rise in the microclimate temperature observed for the mattress topper without phase change material treatment leads to a substantial amount of moisture buildup in the microclimate resulting from an increased perspiration of the skin. The results are summarized in Fig. 9-10.

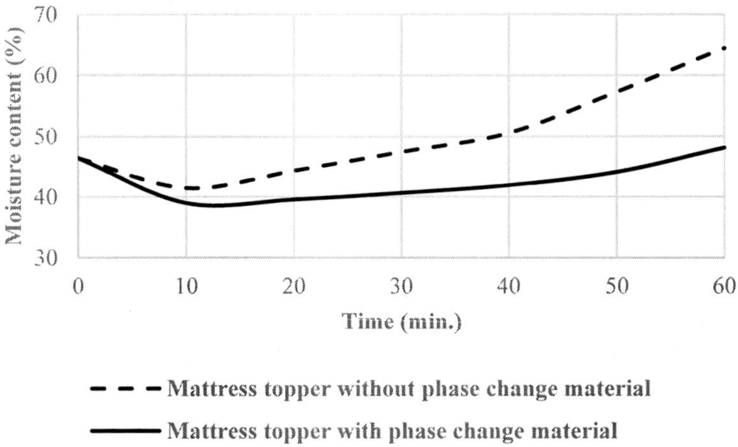

Fig. 9-10: Moisture content in the microclimate.

Without the phase change material treatment, the moisture content in the microclimate exceeded 50 percent after 40 minutes of test. In contrast, with the phase change material treatment, the moisture content in the microclimate remained below 50 percent throughout the entire test.

The test protocol also took into consideration that people are moving during sleep [11]. The human body lies atop a specific area of the mattress only over a period of one to two hours on average. A respective test was performed over a seven-hour period. At the beginning of the test, the test torso form was arranged atop the mattress topper and remained there for one hour. Afterward, the torso form was removed from the mattress and was kept away for another hour. Subsequently, the procedure was repeated at

1.5-hour intervals. Finally, the torso form was placed atop the mattress topper for a two-hour period. The test results are summarized in Fig. 9-11.

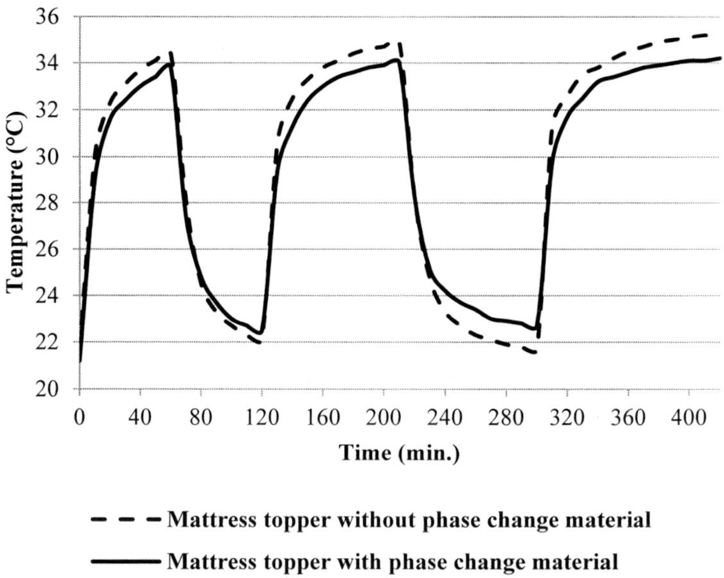

Fig. 9-11: Temperature development in the microclimate over a seven-hour period.

The results shown in Fig. 9-11 indicate that the latent heat absorption of the phase change material can keep the microclimate temperature in the comfort range below 34 °C during various periods such as one hour, 1.5 hours, and 2 hours.

When the mattress topper is uncovered, the phase change material is recharged and able to fulfill its thermal function again. Without the phase change material application, the microclimate temperature drops in the time the mattress is uncovered as well, but always exceeds the comfort range when the torso form, which represents the human body, is lying on the mattress again. The charging and recharging of the phase change material during the night shows the advantage of the thermoregulating feature of the phase change material.

References

1. Pause, Barbara. 2000. "New heat protective gloves with improved thermal capabilities". 2. International Conference on Safety & Protective Fabrics, Arlington, VA, USA, April 26-28, 2000
2. International Organization for Standardization (ISO). 2015. "Clothing for protection against heat and flame – Determination of contact heat transmission through protective clothing or constituent materials; Part 1: Contact heat produced by heating cylinder". ISO Standard No. 12127-1
3. American Society for Testing and Materials (ASTM) International. 2018. "Standard Test Method for Evaluation of Conductive and Compressive Heat Resistance (CCHR)". Standard F 1060-18. Book of Standards, Vol. 11.03
4. Müller, Rainer. 1990. "Kältearbeit in der Nahrungs-und Genussmittelindustrie". Klima am Arbeitsplatz, Schriftenreihe der Bundesanstalt für Arbeitsschutz, TB. 53: 162 –173
5. International Organization for Standardization (ISO). 2008. "Ergonomics of the thermal environment - Cold workplaces - Risk assessment and management". ISO Standard No. 15743
6. Pause, Barbara. 1995. "Development of the first cold protective clothing with microencapsulated PCM". 7th International Techtextil Symposium Frankfurt, Germany, June 19-21, 1995
7. European Committee for Standardization. 2017. "Protective clothing- Ensembles and garments for protection against cold". Standard EN 342
8. American Society for Testing and Materials (ASTM) International. 2020. "Standard Test Methods for Mass per Unit Area (Weight) of Fabric". Standard E D3776/D3776M-20. Book of Standards, Vol. 07.01
9. American Society for Testing and Materials (ASTM) International. 2019. "Standard Test Method for Thickness of Textile Materials". Standard D1777-96. Book of Standards, Vol. 07.01
10. American Society for Testing and Materials (ASTM) International. 2023. "Standard Test Method for Thermal Resistance, Evaporative Resistance, and Total Heat Loss Measurements of Clothing Materials Using a Sweating Hot Plate". Standard F1868-23. Book of Standards, Vol. 11.03
11. Shen, Tsu-Yu, Tan, Su-Huey, Wu, Fong-Gong. 2015. "Comparison of Sleep Quality and Thermal Comfort for Innovative Mattress Design. Procedia Manufacturing. No. 3: 5874-5880,

DOI: 10.1016/j.promfg.2015.07.893

12. Pause, Barbara. 2007. "Application of phase change and shape memory materials in medical textiles". In Smart textiles for medicine and healthcare: Materials, systems, and applications, edited by L. Van Langenhove, 74-87, Woodhead Publishing Ltd.
13. Mecheels, Jürgen. 1991. "Körper-Klima-Kleidung: Grundzüge der Bekleidungsphysiologie". Schiele & Schön Verlag, Berlin, Germany
14. Liu, Jiaping, Liu, Yanfeng, Song, Cong, Zhou, Xiaojun. 2015. "Investigation of Human Thermal Comfort in Sleeping Environments Based on the Effects of Bed Climate". Procedia Engineering. Vol. 121: 1126-1132, DOI: 10.1016/j.proeng.2015.09.118
15. Dongmei, Pan, Mengjie, Song, Ning, Mao, Zhongping, Lin. 2017. "A study on the effects of different bedding systems on thermal comfort – quantifying the sensitivity coefficient used for calculating Predicted Mean Vote (PMV) in sleeping environments". Energy Procedia. Vol. 142: 1939 – 1946, DOI: 10.1016/j. egypro.2017.12.393
16. Aparicio, Inmacculada, Gil-Calvo, Marina, Lucas-Cuevas, Angel G., Perez-Soriano, Pedro, et al. 2017. "Assessment of a mattress with phase change materials using a thermal and perception test". Experimental Thermal and Fluid Science. Vol. 81: 358-363, https://doi.org/10.1016/j.expthermflusci.2016.10.024

CHAPTER TEN

APPLICATIONS OF PHASE CHANGE MATERIAL FOR IMPROVING ENERGY EFFICIENCY

10.1 Principles

Phase change materials are suitable for all kinds of thermal storage applications, such as intermediate or total storage. Thermal storage systems are used, for instance, in conjunction with solar or solar/thermal installations. In addition, phase change materials can be used to regulate the heat flux into and out of a structure, increasing the structure's energy efficiency. However, improving energy efficiency also means keeping a battery's temperature in its optimal operation range.

10.2 Building applications

More than one-third of the overall annual energy consumption of a building is utilized for space heating or cooling. To meet the rising demand in energy supply for residential and commercial buildings, new solutions need to be discovered. Phase change materials are considered as part of the solution. Therefore, several systems based on the application of phase change material have been developed in recent years.

10.2.1 Building materials for exterior or interior use

The use of phase change material as passive energy storage systems in buildings has been envisioned to enhance the thermal mass of the building's envelope, reduce indoor temperature fluctuations, and shift heating and cooling loads away from peak demand times [1].

For instance, existing building materials such as concrete and gypsum plasterboards used for exterior and interior walls and ceilings have been equipped with microencapsulated phase change material operating in a temperature range between approximately 15 °C and 30 °C.

Concrete

In the concrete application, microencapsulated phase change material is mixed into the concrete components replacing a certain amount of its ingredients, i.e., sand or cement. Microencapsulated phase change material has mainly been used in concrete made of Portland cement. Studies conducted by Berardi and Gallardo have shown that the incorporation of microencapsulated phase change material into concrete reduces the peak temperature of hydration by approximately 3 degrees Celsius by absorbing the heat released in the cement's hydration process. This could cause a delay in the time required to reach the peak hydration temperature and interfere with the hydration reaction. Investigations have also pointed out that the workability and the flow values of concrete mixtures, as well as the compressive strength of the concrete, are significantly reduced when microencapsulated phase change materials are added to the system [2, 3].

Mechanical strength and workability requirements restrict the maximum amount of microencapsulated phase change material that can be incorporated into concrete. Based on these requirements only 3 wt.% to 5wt.% of microencapsulated phase change material can be applied to the concrete to achieve the minimum compressive strength [2].

A concrete block with a size of 0.2 meters by 0.2 meters by 0.4 meters weighs about 17 kilograms. Based on an application of 3 wt.% to 5 wt.%, about 500 grams to 850 grams of microencapsulated phase change material could be applied to such a concrete block. The microcapsules possess a latent heat storage capacity of about 105 J/g. Hence, a latent heat storage capacity of the microencapsulated phase change material incorporated into the concrete block between 53 kJ and 90 kJ is obtained. The specific heat capacity of concrete is 1.05 kJ/kg K. Therefore, a concrete block with a weight of 17 kilograms absorbs a sensible heat of about 18 kJ while its temperature rises by one degree. This leads to a sensible heat absorption of about 180 kJ over the temperature range in which the latent heat absorption of the microencapsulated phase change material occurs. Thus, the total heat absorption of the concrete block is increased by 30 percent to 50 percent when the latent heat absorption of the phase change material takes place.

The microencapsulated phase change material at a concentration of 5 wt.% substantially decreases the thermal conductivity of the concrete. Berardi and Gallardo's studies showed that the thermal conductivity was reduced by 25 percent to 50 percent. This reduction in thermal conductivity is advantageous in terms of thermal insulation but also significantly influences the charging and recharging of the microencapsulated phase change material integrated into the concrete [2].

Gypsum plasterboards

According to a study conducted by Bake et al., gypsum plasterboards with microencapsulated phase change material were manufactured and tested. To make gypsum plasterboards, multi-finished gypsum was mixed with water and microencapsulated phase change material. The mixture was poured into a form and the material was dried for 28 days. It was found in the tests that a quantity of 15 wt.% of microencapsulated phase change material can be incorporated into a gypsum matrix at maximum [4]. Gypsum plasterboards were made in a size of 0.95 meters by 1 meter by 0.01 meters. The weight of the manufactured gypsum plasterboard totals about 15 kilograms. A microcapsule contribution of 15 wt.% in the gypsum plasterboard results in a weight of 2.25 kilograms. Under consideration of a latent heat storage capacity of the microencapsulated phase change material of about 105 J/g, a total latent heat storage capacity of the manufactured gypsum plasterboard configuration of about 236 kJ was obtained [5].

The specific heat capacity of gypsum plasterboards is 0.84 kJ/kg K. Therefore, the gypsum plasterboard with a weight of 15 kilograms absorbs only a sensible heat of about 13 kJ while its temperature rises by one degree. This leads to a sensible heat absorption of about 130 kJ over the temperature range in which the latent heat absorption of the microencapsulated phase change material occurs. Hence, the total heat absorption of the gypsum plasterboard is nearly doubled when the latent heat absorption of the phase change material takes place.

The addition of the microencapsulated phase change material reduces the thermal conductivity of the gypsum plasterboards by about 27 percent. This enhances the thermal insulation of the gypsum plasterboards. It is advantageous in terms of heat transfer, but the charging and recharging of the phase change material could be delayed based on this thermal conductivity reduction. Furthermore, it was determined in tests that the gypsum plasterboard's bulk density and compression strength were reduced by the application of microencapsulated phase change material [5].

Ceiling tiles

In an interior room, the warm air usually rises to the top and increases the temperature of the ceiling. The convection of air and the formation of drafts in the room, as well as a high temperature gradient between a cold floor and a warm ceiling, leads to an experience of thermal discomfort by the occupants of the room. The temperature gradient between the floor and the ceiling should preferably be less than 3 degrees Celsius to ensure a thermal comfort sensation inside the room [6].

Phase change material incorporated into ceiling tiles will absorb excessive heat rising to the ceiling as soon as the trigger temperature of the phase change is obtained. The heat will be absorbed as latent heat while the temperature of the phase change material and the surrounding ceiling tiles remains constant. As a result, the temperature differences between the floor and the ceiling will be kept low and consistent during the day.

The phase change material applied to a ceiling tile should absorb latent heat when its temperature rises above 24 °C and should release its stored latent heat when its temperature drops below 21 °C. A salt hydrate mixture (salt eutectic) operating in this temperature range was determined to be suitable for the ceiling tile application. The salt hydrate mixture possesses a latent heat storage capacity of 127 J/g.

A common ceiling tile made of a mineral fiber board possesses a size of 0.625 meters by 0.625 meters and has a thickness of about 0.015 meters. The salt hydrate mixture was filled into cavities inside the ceiling tile in a quantity of 1.2 kilograms. Based on this quantity, a latent heat storage capacity of one ceiling tile of about 152 kJ has been obtained leading to a latent heat storage capacity of the ceiling of 56 kJ/m^2. The latent heat storage capacity realized by the salt hydrate mixture incorporated in the ceiling tiles ensures an efficient latent heat absorption which, in turn, keeps the indoor climate in the comfort range. The only issue related to such a ceiling configuration is recharging the salt hydrate mixture by releasing its stored latent heat. The salt hydrate mixture might show a supercooling behavior and its latent heat storage capacity could decrease over time.

Wall coverings

Wall coverings play an important role in thermal interactions between the walls, the ceiling, and the room's interior. The walls release radiant heat into the room and receive radiant heat from the room's interior parts. Furthermore, the heat flux from the floor to the ceiling interacts with the wall covering in the form of convective heat. If phase change material is part of the wall covering system, it can play an active role in the temperature control of the interior by absorbing excessive heat or releasing heat, if needed, providing a thermoregulating effect. This leads to a substantial comfort improvement [6]. Because the phase change material is located on the wall's surface, there is no thermal barrier, such as surrounding material with low thermal conductivity to overcome. Therefore, the charging and recharging of the phase change material will not be as difficult as it is, for instance, in walls made of concrete.

Microencapsulated phase change material can be incorporated into clay plaster mortar in a concentration of up to 15 wt.% based on the weight of the dried mortar. In a study by Sevilla et al., clay mortar samples were manufactured by mixing clay, water, and phase change material microcapsules together, and curing them within 14 days. Test results received for the mortar samples in the study indicate that the density and the workability decrease with a rise in the microcapsule content. The addition of the microcapsules did not influence the thermal conductivity. For a phase change material microcapsule concentration of 15 wt.% in the clay plaster mortar, a latent heat storage capacity of 10.6 J/g was determined in the research study conducted by Sevilla et al. [7].

Another possibility to apply phase change material to interior walls is the wall covering assembly disclosed in US Patent no. 7,241,509, made of a common wallpaper or fabric that is coated on the backside with an acrylic compound containing phase change material [8]. A second coat of a liquid ceramic compound is applied on top of the acrylic coating layer to meet fire resistance requirements. The ceramic coat also provides a thermal resistance which prevents heat penetration into the adjacent wall. The wall covering assembly is applied to the wall in the manner of a wallpaper.

The phase change material applied to the wallcovering absorbs external heat penetrating through the windows during the day as well as internal heat generated by heat sources inside the room. The phase change material releases its stored latent heat overnight in a reverse heat flux through the windows. It has been determined that the phase change material should absorb latent heat when its temperature rises above 22 °C and should release its stored latent heat when its temperature drops below 18°C. Heptadecane paraffin operates in this temperature range. In a study, microencapsulated heptadecane was added to the acrylic coating compound in a quantity of about 130 g/m^2. Based on a latent heat storage capacity of 170 J/g of the selected microencapsulated heptadecane, a latent heat storage capacity of 22 kJ/m^2 was obtained. The phase change material application in the wall covering will lead to increased thermal comfort inside buildings as well as energy savings.[8].

10.2.2 Roofing systems

Considering the thermal performance of an entire structure, the roof is a significant weak point. During the day, most of the heat provided by solar radiation penetrates through the roof into the building. On hot and cloudless days, the living area below the roof often overheats which leads to a substantial decrease in thermal comfort. Then, overnight, most of the heat stored inside the building during the day is lost through the roof. On cold days, this significant heat loss through the roof has a large influence on the heating demands of the structure [9, 10].

Based on thermal tests in common roof configurations and the utilization of a computer modelling procedure, it was determined that phase change material should be arranged in the following two locations in a common roof structure:

- In the upper part of the roof below the shingles.
- In the lower part of the roof between the thermal insulation and the drywall.

Upper roof application

The main purpose of the phase change material application in the upper part of the roof below the shingles is to decrease the heat transfer into the structure on hot and cloudless days. The phase change material situated in this location will absorb the heat generated by solar irradiation penetrating through the shingles. As soon as a given trigger temperature is obtained, the phase change material will absorb latent heat, and its temperature will remain nearly constant. This limits the heat flux in the structure. The heat flux, which would normally rise during the day, will remain at the low level experienced in the early morning hours. The latent heat absorbed by the phase change material during the day is released as soon as the ambient temperature drops in the early evening hours and the trigger temperature for the latent heat emission by the phase change material is obtained. Because of the thin shingles and the proximity of the phase change material location to the surroundings, a complete recharge of the phase change material overnight can be expected.

To select an appropriate phase change material for the upper roof application, temperature measurements and computer modelling procedures were carried out. The heat gain through common roof structures under various ambient conditions, as well as temperature developments inside model buildings, were studied. Based on the investigations, it was determined that the phase change material should absorb latent heat in a temperature range between 50 °C and 60 °C. A phase change material operating in this temperature range was selected, which is non-combustible and possesses a latent heat storage capacity of 200 J/g. The phase change material has been applied to an acrylic coating compound in a quantity of 50wt.% which leads to a latent heat storage capacity of about 50 kJ/m^2. The coating compound with the phase change material incorporated wherein is spray-coated onto the backside of the shingles. The approximately 2-millimeter-thick coating layer increases the weight of the shingles by about 14 percent. Furthermore, the coating enhances the thermal resistance of the shingles by about 20 percent.

The shingles treated with phase change material were tested on a hot and cloudless day. The ambient temperature rose from 25 °C in the morning to 38 °C in the afternoon. Sample shingles with and without the phase change material treatment were tested side by side in a roof configuration facing south. The temperatures on the backside of the shingles were measured between 9 a.m. and 5 p.m. (17). The test results are shown in Fig. 10-1.

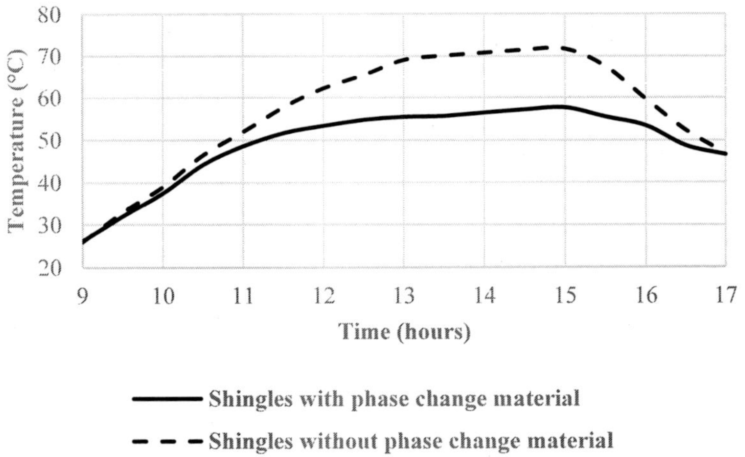

Fig. 10-1: Temperatures on the backside of the shingles received in the field test.

As indicated in Fig. 10-1, the temperature on the backside of the shingles rises continuously in the late morning and early afternoon hours and reaches its maximum at about 3 p.m. (15) in the afternoon under daylight savings time conditions. At the end of the heat gain at 3 p.m. (15), a high temperature of about 72 °C was measured behind the shingles without phase change material treatment.

The latent heat absorption of the phase change material included in the backside coating of the shingles reduces the shingle temperature by about 14 degrees Celsius at maximum. The latent heat absorption by the phase change material was initiated at 11 a.m. and lasted until 3 p.m. (15), i.e., over four hours. After 3 p.m. (15), the sun irradiation no longer hit the south-facing shingles; Therefore, the shingles cooled down, and the temperature on their backside dropped. The recharge of the phase change material was completed at about 5 p.m. (17).

With a rise in the shingle temperature during the day, the temperature gradient between the roof, the structure's attic, and the living space underneath the attic increases, leading to a rise in the heat flux into the structure. In contrast, the latent heat absorption of the phase change material prevents a substantial gain in the shingle's temperature, limiting the heat flux into the building.

Fig. 10-2 shows the heat flux into the structure during the day between 9 a.m. and 5 p.m. (17) for two similar roofing systems that differ only by the application of shingles with and without the phase change material treatment. The calculated results are based on temperature measurements received in the field test and thermal insulation values of the roof components. The columns in Fig. 10-2 represent mean values of the heat flux through the roof into the building over one-hour periods.

As the results in Fig. 10-2 indicate, for a common roofing system without phase change material treatment, the heat flux through the roof constantly rises during the day. It reaches its peak values in the early afternoon between 1 p.m. (13) and 3 p.m. (15) under daylight savings time conditions. In contrast, the latent heat absorption of the phase change material prevents a significant rise in the heat flux through the roof, even under increasing sun exposure in the late morning and early afternoon hours.

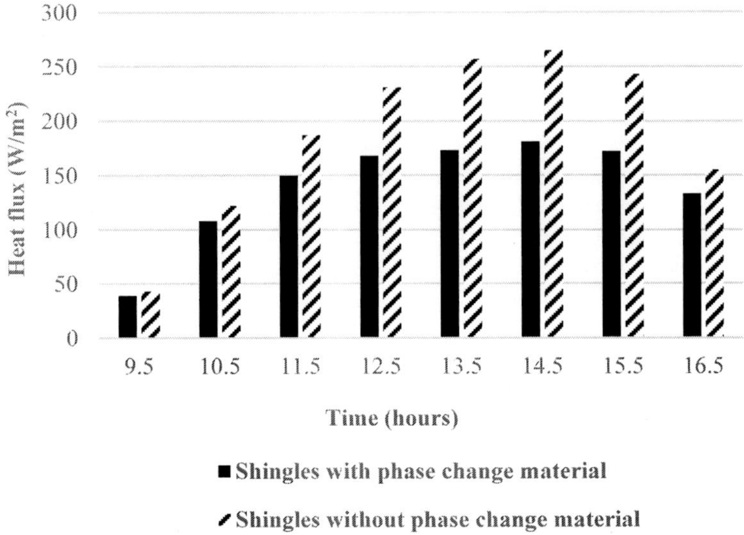

Fig. 10-2: Heat flux through the roof into the structure during the day.

For instance, the heat flux into the structure was decreased on average by about 32 percent under the highest sun exposure between 1 p.m. (13) and 3 p.m. (15) due to the latent heat absorption of the phase change material. Over the course of the day, the heat flux through the roofing system was reduced by about 22 percent by the latent heat absorption of the phase change material.

Based on the results of the field tests, computer modelling procedures were performed to study roofing systems of residential and commercial buildings where shingles were treated with phase change material. As a result of the phase change material treatment, the interior temperatures could be stabilized without a high cooling load of an air-conditioning system. The calculation indicates that the downsizing, and the limited use of an air-conditioning system could lead to energy savings of up to 25 percent. In addition, peak energy demands, especially during hot weather conditions, could be avoided in this way.

Lower roof application

The phase change material applied in the lower part of the roof structure just above the drywall provides temporary additional thermal insulation on cold nights. In this case, the phase change material will absorb and store the heat generated in the living space below the roof during the day after penetrating through the drywall as soon as the trigger temperature for the latent heat absorption is obtained.

When the temperature of the phase change material drops overnight below the trigger point, the latent heat stored in the phase change material will be released into the insulation layer. The heat will stay in the insulation layer for a while and slowly dissipate, limiting the heat flux through the roof components into the environment overnight.

In addition to this heat flux control through the roof overnight, the temperature of the ceiling atop the living space will not rise during the day due to temperature consistency caused by the latent heat absorption of the phase change material. This reduces the temperature gradient between the living space's floor and ceiling, enhancing the thermal comfort inside the room [11].

10.2.3 Façade systems

A façade system is a structural and aesthetic component of a building. It gives the building a unique character and protects against the elements. Façade systems are typically made of glass, aluminum, composite materials, or wood panels. Nowadays, textile membranes are also used [12, 13].

Textile membranes are made, for instance, of polyester or fiberglass fabrics coated either with PVC, PTFE, or silicone rubber compounds. Such membranes are commonly used for roofing structures, stadium covers, or canopies. The roof of the Denver International Airport is made of a PTFE-coated fiberglass membrane, as shown in Fig. 10-3.

Textile membranes can be treated with phase change material to obtain thermoregulating properties. Fiberglass fabrics coated with a silicone rubber compound are the most suitable membrane configurations for this application. To accommodate the climatic conditions in various geographical regions, different phase change materials have been selected that are compatible with the silicone rubber compound. Non-combustible salt hydrates are the preferable candidates to meet fire-resistant requirements.

Fig. 10-3: Textile membrane roof of the Denver International Airport.

First, the phase change material is integrated and cross-linked into a two-component silicone rubber coating compound. The compound is then topically applied to both sides of a fiberglass fabric by a knife-over-roll coating procedure [14].

The membrane materials with phase change material treatment can be made in different versions, i.e., different weights and thicknesses. In one version, a 0.5-millimeter-thick fiberglass fabric was coated on both sides with a 0.5-millimeter-thick silicone rubber compound containing a salt hydrate in a concentration of 40 wt.% [15, 16, 17, 18]. Some characteristics of the silicone rubber-coated fiberglass membrane treated with the selected salt hydrate phase change material are summarized in Table 10 -1.

Weight/unit area, g/m^2	1,440
Thickness, mm	1.5
Temperature range of the phase change material's latent heat absorption, °C	25 – 40
Temperature range of the phase change material's latent heat release, °C	15 – 25
Membrane's latent heat storage capacity, kJ/m^2	130

Table 10-1: Characteristics of the membrane fabric treated with a salt hydrate phase change material.

The membrane treated with phase change material can be applied to a façade cladding system in new construction or used for retrofitting an existing building. The membrane can be arranged as a façade envelope supported by a frame made of steel or aluminum elements [19]. Fig. 10-4 shows a possible design of such a façade system.

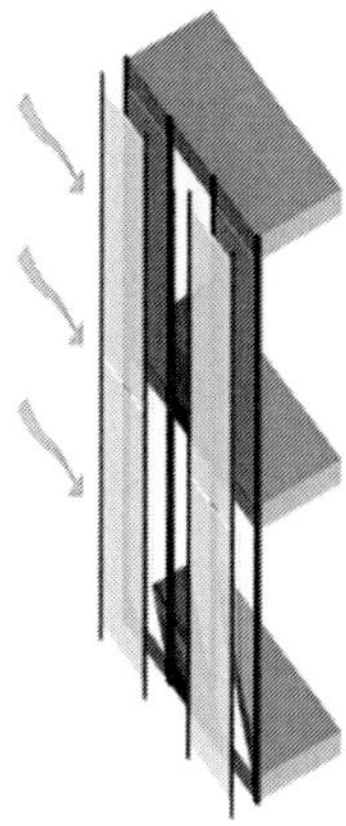

Fig. 10-4: Design of a façade system.

The membrane materials with the phase change material treatment possess two unique features:

- Adaptive heat transfer properties.
- Adaptive light transmission properties.

Both features are beneficiary for an application in a façade system.

Adaptive heat transfer properties

Without the application of phase change material in the membrane material, the solar irradiation that hits the facade during the day results in a heat gain inside the structure. This may contribute to an overheating of the structure, especially in the afternoon and evening hours on hot and cloudless days. In contrast, the phase change material incorporated into the coating of the fiberglass fabric controls and limits the heat flux through the façade into the structure, mitigating overheating of the building during the day. This heat flux control is triggered by temperature. Therefore, membrane materials treated with phase change material are considered "smart" materials [17, 18].

The membrane temperature increases as the sun rises in the morning hours, and the heat transfer through the façade into the building intensifies. As soon as the phase change material embedded into the membrane material reaches its trigger temperature in this heating process, the phase change material starts to absorb latent heat. This leads to a delay in the rise in the temperature of the membrane treated with phase change material. Test results of the temperature development on the backside of the membranes utilized in the façade configuration received in a test on a hot and cloudless day are summarized in Fig.10-5. In the test, the membrane material characterized in Table 10-1 was compared to a common silicone-coated fiberglass membrane without phase change material treatment with a similar weight and thickness. The two membranes were tested side by side, facing south.

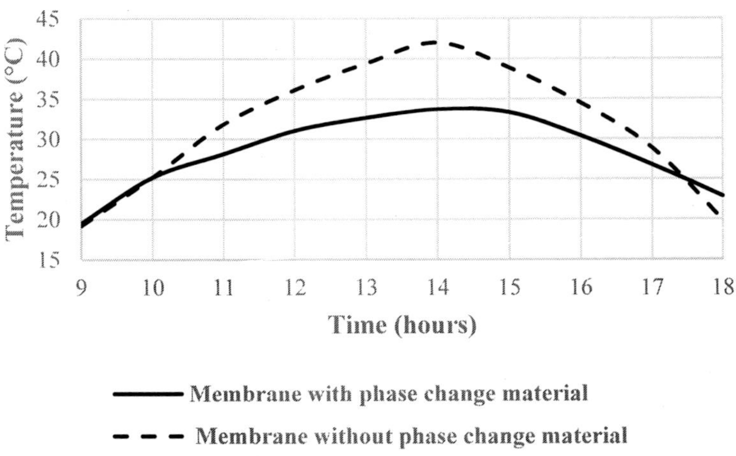

Fig. 10-5: Temperature development on the backside of the membranes.

As the results in Fig. 10-5 indicate, the latent heat absorption of the phase change material limits the rise in the membrane's temperature during the day. On the other side, the temperature of the common membrane rises substantially over the course of the morning and early afternoon hours and reaches its maximum at 2 p.m. (14) under daylight savings time conditions. A temperature difference of 8 degrees C between the two membranes was obtained at 2 p.m. (14) in the afternoon.

With a rise in the membrane's temperature during the day, the heat transfer through the façade into the structure increases. On the other hand, the latent heat absorption of the phase change material incorporated into the membrane prevents a substantial increase in the membrane's temperature and, therefore, limits the heat transfer through the façade into the building.

Fig. 10-6 shows the heat flux through the facade for the two membrane configurations between 9 a.m. and 6 p.m. (18). The columns in Fig. 10-6 represent mean heat flux values through the facades over one-hour periods.

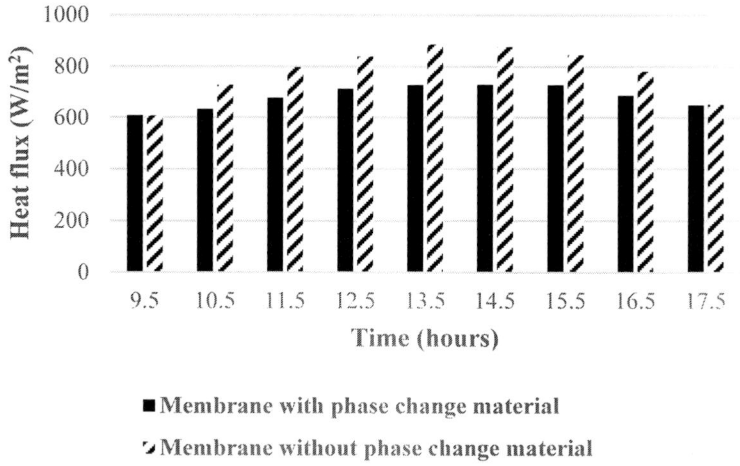

Fig: 10-6: Heat flux through the façade into the structure.

The results summarized in Fig. 10-6 show that the heat flux through the façade into the building is substantially reduced in the early afternoon by the application of the phase change material in the membrane. For instance, a difference in the heat flux through the two membrane facades of about 18 percent is obtained in the early afternoon hours between 1 p.m. (13) and 2 p.m. (14). Over the course of the day, the heat flux through the façade was reduced by about 15 percent by the latent heat absorption of the phase change material.

The heat flux reduction is lower for the façade application of the phase change material than for the upper roof application of the phase change material despite the higher latent heat storage capacity. In the façade application, the membrane with the phase change material is directly hit with solar irradiation. At the same time, the shingles provide a thermal buffer before the phase change material itself is exposed to solar irradiation.

Adaptive light transmission properties

The translucency of the silicone rubber-coated fiberglass membrane equipped with phase change material characterized in Table 10-1 exceeds that of common membrane materials such as PVC-coated Polyester and PTFE-coated fiberglass [20].

In addition, the silicone rubber-coated fiberglass membrane with phase change material treatment changes its light transmission over the course of the day. Most of the phase change materials used in the silicone-coated fiberglass membranes are opaque in their solid state and transparent in their liquid state. The silicone rubber coating compounds utilized in membrane material configurations are primarily transparent. When the phase change material absorbs latent heat and melts, its appearance changes gradually from opaque to transparent. At the same time, the entire membrane material becomes increasingly more translucent. The light transmission change is triggered by temperature. Therefore, the membrane material treated with phase change material is also considered a "smart" material in this regard [20].

The difference in light transmission between the two states (solid and liquid) of the phase change material incorporated into the silicone rubber compound that is coated onto the fiberglass fabric totals up to 27 percent. At night, the phase change material is opaque in its solid state. The light transmission through the silicone rubber-coated fiberglass membrane equipped with the phase change material in this state is only 26 percent, which aids privacy. Once the phase change material starts to absorb latent heat in the morning, the membrane becomes more translucent. This leads to an increase in the transfer of light through the membrane façade into the building. When all the phase change material is melted, the translucency of the silicone rubber-coated fiberglass membrane treated with phase change material totals about 53 percent. As a result, the use of artificial light in the interior throughout the day can be minimized. The transfer of natural light into the building is important for the productivity and well-being of the people who work and live there. In addition, the energy and cost savings resulting from eliminating or reducing artificial lighting during the day must be considered [21].

Fig. 10-7 summarizes the results received for the change in the translucency of the façade made of silicone rubber-coated fiberglass with the phase change material application during the day. The translucency of a common silicone rubber-coated fiberglass membrane is used for comparison.

While the translucency of the common silicone rubber-coated fiberglass membrane without phase change material remains constant at about 21 percent, the translucency of the membrane with phase change material treatment increases substantially in the early afternoon.

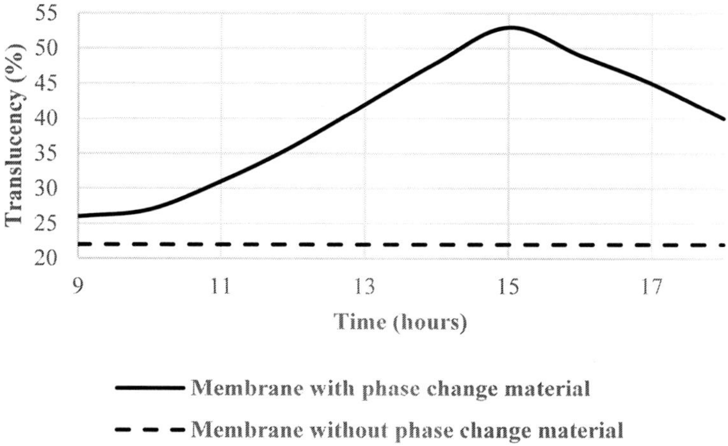

Fig. 10-7: Changes in translucency during the day.

The melting process of the phase change material leads to a curious and somewhat counterintuitive effect: the infrared part of the spectrum of the sun radiation is effectively filtered out while the transmission of the visible part of the light spectrum increases at the same time. In façade applications, this can eliminate the need for costly and often complex shading systems.

Controversially, when the cooler outside temperature overnight results in the solidification of the phase change material, the light transmission through the membrane is reduced. This, in turn, increases the membrane's capacity to reflect light from interior sources. As a result, the efficiency of interior lighting and especially the efficiency of indirect lighting fixtures directed toward the membrane, is increased.

10.3 Solar application

Phase change material can be used to enhance the efficiency of photovoltaic (PV) cells and photovoltaic (PV) modules. The main part of a solar cell, the semiconductor is very sensitive to temperature changes. PV cells are tested for their efficiency at 25 °C, which is the cell's temperature under Standard Test Conditions (STC). A rise in the cell's temperature above 25 °C leads to power losses. In general, there is a power loss of 1 percent to be considered for every 2 °C temperature increase. During operation, the temperature on top of the PV cell easily reaches 65°C on hot and cloudless days. This temperature increase results in about a 20 percent loss in efficiency [22, 23, 24].

Phase change material can be used to draw heat away from the PV cell and absorb this excessive heat while keeping the PV cells at a lower temperature during the day. This will enhance the PV cells' efficiency and lead to increased power output. The latent heat stored in the phase change material, in turn, is used for heating water [25, 26].

10.3.1 Textile-based hybrid PV/T collector

A textile-based hybrid photovoltaic/thermal (PV/T) collector was developed, in which phase changer material is utilized as an intermediate thermal storage medium. A picture of one version of the textile-based hybrid PV/T collector is shown in Fig. 10-8.

Fig. 10-8: Textile-based hybrid PV/T collector.

The collector presented in Fig. 10-8 consists of a spacer fabric that integrates capillary pipes. The front side of the spacer fabric faces the PV cells. This side is coated with a compound that contains phase change material. The PV cells are laminated onto the surface of the coating layer. The coating compound penetrates the fabric during the coating procedure and encloses the capillary pipes. The backside of the spacer fabric is covered with a protective coating to withstand the elements.

The PV cells located atop the collector convert the solar radiation into electricity. As soon as the temperature behind the PV cells reaches a given trigger temperature, the heat is drawn away from the PV cells and absorbed in the form of latent heat by the phase change material. In this way, the PV cells are cooled and can operate at a lower temperature, staying at their highest possible efficiency and maximum electrical output.

The textile-based hybrid PV/T collector is lightweight, thin, flexible, and weatherproof. It fits any roof shape and can be used on roofs of residential and commercial buildings, swimming pool covers, and greenhouse covers. Another application could be emergency shelters and temporary hospitals set up in disaster areas. The system would provide shelter, electricity, and hot water [27].

10.3.2 Hot water generation system

The absorbed latent heat is stored in the phase change material and is released when water is pumped through the capillary pipes. This process recharges the phase change material for the next cycle of heat absorption. The system used for the hot water generation is shown in Fig. 10-9.

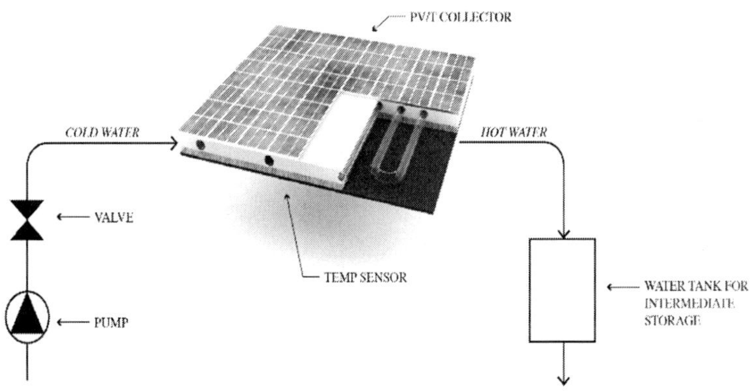

Fig. 10-9.: Hot water generation system.

Temperature measurements performed by a temperature sensor deliver an indication of when the phase change material is ready for a recharge. At this point, the phase change material has stopped absorbing latent heat and its temperature starts to rise. The temperature sensor provides a signal to the valve which normally shuts off the water supply. The valve opens, and cold water with a temperature of about 18 °C is pumped into the capillary pipes inside the collector [27, 28].

Because of the high gradient between the temperature of the phase change material and the temperature of the water at the beginning of the recharge cycle, the latent heat stored in the phase change material is quickly transferred to the water, and the water is heated. The water which is flowing out of the collector has a temperature of about 42 °C.

Water flow through the capillary pipes can be initiated automatically using a thermostat or manually. The hot water can be utilized instantaneously or stored in an intermediate hot water tank. This procedure eliminates the need for a heat exchanger and its components, as well as the need for a large storage tank which, in turn, reduces costs and saves space.

10.3.3 Tests of the hybrid PV/T system

The phase change material utilized in the PV/T collector absorbs latent heat between 25 °C and 40 °C. The selected phase change material possesses a high latent heat storage capacity of about 295 J/g. Applied to the coating compound located behind the PV cells in a concentration of 50 wt.%, a latent heat storage capacity of about 580 kJ/m^2 is obtained.

Cooling of the PV cells

The main purpose of utilizing phase change material in the PV/T collector is the cooling of the PV cells to ensure that they remain in the optimal temperature range during the window of operation. Field tests were carried out in which the temperature on the surface of the PV cells was measured. In the test, collectors with and without the phase change material application were placed side by side and the temperature development at the PV cells was monitored during the day. The test results are summarized in Fig.10-10.

The results summarized in Fig.10-10 indicated that the temperature at the PV cell without the phase change material application increased by 45 degrees C between 9 a.m. and 2 p.m. (14). The latent heat absorption of the phase change material starting at a trigger temperature of 25 °C prevents such a temperature increase. The temperature on the PV cell treated with phase change material increased only by 22 degrees C over the same period.

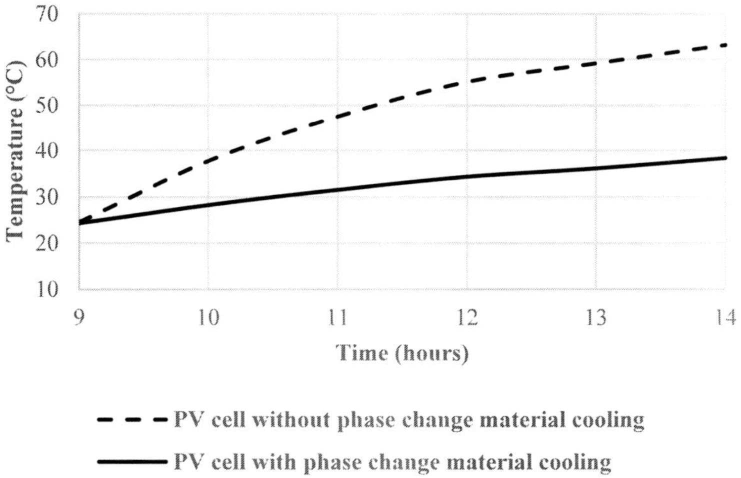

Fig. 10-10: Temperature development at the surface of the PV cells with and without cooling by phase change material.

Power generation of the PV cells

The cooling of the PV cells by the phase change material leads to a substantial improvement in the power output, as shown in Fig. 10-11. The power output from the PV cells of the collectors with and without phase change material was measured simultaneously with the temperature measurements.

As shown in Fig.10-11, the cooling of the PV cell provided by the latent heat absorption of the phase change material leads to a reduction of the power output by only 5 Watts or 7 percent during the temperature increase between 9 a.m. and 2 p.m. (14). In contrast, without the cooling feature provided by the phase change material, the power output of the same PV cell dropped by 15 Watts or 19 percent due to its overheating in the same time frame.

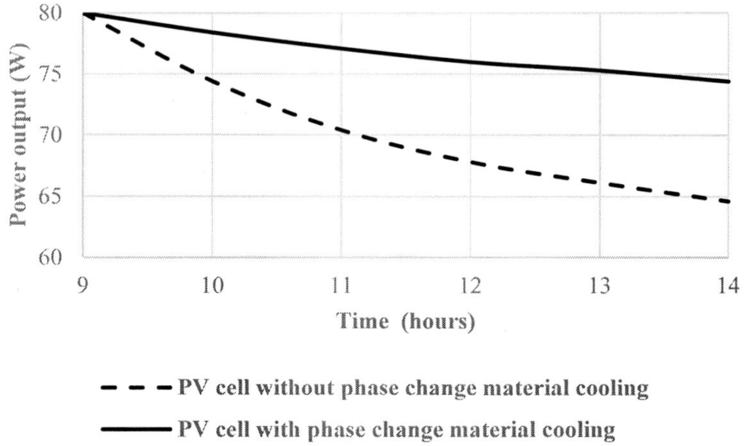

Fig. 10-11: Power output of the PV cells with and without cooling by phase change material.

Hot water generation

The latent heat storage capacity of about 580 kJ/m^2 allows for heating about 6 liters of water from 18 °C to 42 °C, based on a collector size of one square meter in a single recharge. Multiple recharges are usually carried out over the course of the day. The textile-based hybrid PV/T collector size between 8 m^2 and 10 m^2 seems sufficient for use in a single-family residence. The water storage tank should have a capacity of about 150 liters to 250 liters.

10.4 Automotive applications

Phase change material can be utilized for the thermal management of automotive passenger compartments and batteries in electric vehicles.

10.4.1 Temperature management of electric vehicle batteries

The optimal battery capacity used in electric vehicles depends mainly on the battery's temperature. Specifically, the optimal capacity of batteries used in electric vehicles is only established in a temperature range between 20°C and 30 °C. Under these conditions, the maximum driving distance given by the manufacturer is obtained.

Cold weather reduces the capacity of an electric vehicle battery by as much as 40 percent. When the temperature drops below 15 °C, the electrolyte fluid inside the lithium-ion battery pack becomes more viscous. The increased viscosity slows down the chemical reaction responsible for the transfer of the electrons. This impacts not only the maximum driving distance of the vehicle under low ambient temperatures but also the duration of the battery's recharge.

High temperatures, on the other hand, accelerate the battery's aging. The immediate effects are reduced driving distance and acceleration. If the temperature increases above 40 °C, serious and irreversible damage to the battery is often obtained. At temperatures over 80°C, a thermal runaway can occur, in which the battery can catch fire.

It also needs to be considered that in cold and hot weather conditions, the battery power is used not only for driving the vehicle but also for heating or cooling the passenger compartment. These tasks are further reducing the battery's capacity [29].

Conventional cooling solutions for batteries in electric vehicles rely on air or liquids applied from the outside and transported via channels to the inside of the battery pack. Thermoelectric coolers, as another solution to cool batteries of electric vehicles, are based on the Peltier effect where voltage is converted into a temperature difference. However, each of these cooling solutions requires power to operate, which is taken from the battery's capacity [30].

A solution to mitigate the battery temperature in both directions can be obtained by the application of phase change material. A mixture of phase change materials is embedded into a flexible polymeric film from which sheets or sleeves are made. The sheets are inserted between the battery modules as it is shown in Fig. 10-12. If applied, the sleeves cover the entire battery pack [31].

One of the phase change materials absorbs latent heat when the temperature rises above 30 °C, and a second phase change material releases latent heat when the temperature drops below 20°C. Both phase change materials interact with each other in the temperature range between 20°C and 30 °C.

Fig.10-12: Sheets with phase change material inserted between the modules of the battery in an electric vehicle.

The phase change material, which absorbs latent heat above 30 °C, prevents overheating of the battery under high ambient temperatures of up to 40 °C during a certain period, leading to an increased battery capacity and an improved driving distance. The latent heat absorption of this phase change material operating in the higher temperature range can also prevent, to a certain degree, an overheating of the battery during charging. However, a thermal runaway and burn down of Lithium-ion batteries utilized in electric vehicles cannot be handled and prevented by the selected phase change material and its specific application. Other solutions need to be discovered in this regard [32].

On the other hand, the phase change material which releases its stored latent heat at temperatures below 20 °C will keep the battery's temperature in the optimal range for an extended period under ambient temperatures between 10°C and 20 °C preventing a substantial decrease in the battery's capacity which, in turn, ensures that the driving distance will not be significantly reduced. However, this phase change material does not operate below 10°C. Therefore, the problems electric vehicle batteries face at very low ambient temperatures cannot be avoided by this solution.

10.4.2 Thermal control of the passenger compartment

Especially on hot days when a car is parked outside in direct sun exposure, its passenger compartment gets overheated already in a short period. When the car is then later driven away from the parking lot, air-conditioning is used to cool the passenger compartment down. However, the utilization of an air-conditioning system requires a lot of energy. This can be especially problematic for electric vehicles because the energy used for the air-conditioning reduces the battery's capacity, which, in turn, decreases the driving range.

Most of the heat, which increases the temperature inside the passenger compartment during parking penetrates through the metal roof. The heat is absorbed by the headliner and further radiated into the cabin.

Phase change material can be used to control the heat transfer through the roof and the headliner into the cabin. In tests, a phase change material has been utilized in the headliner which absorbs latent heat when its temperature rises above 43 °C. It possesses a latent heat storage capacity of about 180J/g. The phase change material has been integrated into an acrylic coating compound in a concentration of 50 wt. % leading the latent heat storage capacity of about 54 kJ. It has been applied to the top of the headliner in the form of a spray coating procedure.

The absorbed latent heat is stored in the phase change material keeping the headliner's temperature at a lower level. As a result, a much smaller amount of heat is radiated into the cabin during this parking situation. The stored latent heat is released through the metal roof into the environment as soon as the car is driven again. The air movement during driving cools the metal roof which, in turn, creates a temperature gradient between the phase change material layer and the top of the metal roof and forces the phase change material to release its stored latent heat through the roof as soon as the temperature drops below 41 °C. This cooling solution does not require any external energy and is maintenance-free. The solution is durable over the lifecycle of the vehicle.

Thermal performance tests

Headliners with and without the phase change material treatment were tested in a two-hour parking situation where the test vehicles were parked side by side in direct sun exposure. The headliners utilized in the tests consist of a décor cover fabric, an intermediate foam layer, and a nonwoven backing material. They cover an area of 1.8 m^2 and possess a basic weight of 2 kilograms. The phase change material coating is spread over the surface of the nonwoven backing material covering the entire area of the headliner. The coating increases the weight of the headliner by about 15 percent. The temperature development on top of the two headliners during the two-hour parking period is shown in Fig. 10-13.

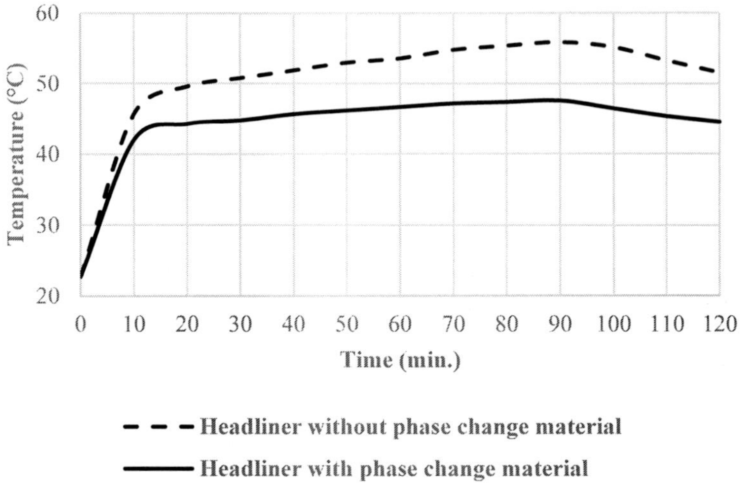

Fig. 10-13: Temperature development at the top of the headliners during parking.

The results summarized in Fig. 10-13 indicate that the latent heat absorption of the phase change material substantially reduces the temperature at the top of the headliner. A temperature difference of a maximum of 8 degrees Celsius has been received in the test. After 90 minutes into the test the temperature on top of both headliners decreases because of the change in the angle in which the sun irradiation hits the metal roof.

The lower temperature caused by the latent heat absorption of the phase change material has a significant influence on the heat flux through the headliner into the passenger compartment. The heat flux has been calculated based on the temperatures on top and at the bottom of the headliner at given times, and the thermal resistance values of the headliner components. The results are shown in Fig. 10-14. On average, the heat flux through the headliner into the passenger compartment has been reduced by about 18 percent due to the phase change material application.

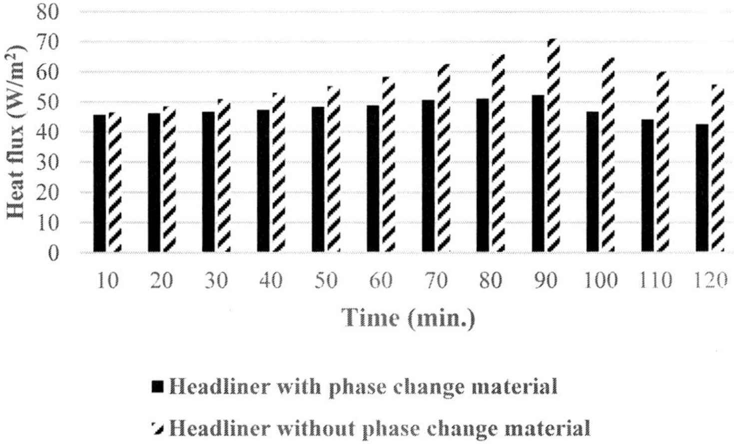

Fig. 10-14: Heat flux through the headliner into the passenger compartment during parking.

The decrease in the heat flux through the headliner due to the latent heat absorption of the phase change material contributes to lower air temperatures in the passenger compartment, as shown in Fig. 10-15.

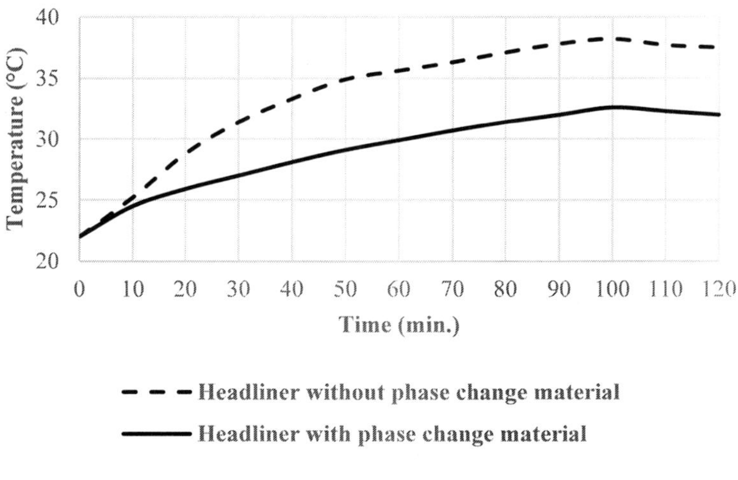

Fig. 10-15: Temperature development inside the passenger compartment during parking.

The air temperatures were measured simultaneously with the temperature on top and at the bottom of the headliner. The air temperatures inside the passenger compartment are also influenced by the heat flux through the windshield and the windows during the parking situation. However, the decrease in the air temperature is due to the reduction in the heat flux through the roof, and hence, the headliner has a significant influence on the cabin's thermal comfort experienced when the vehicle is occupied again after the parking period.

The lower air temperature at the end of the parking period received for the headliner configuration with the phase change material application not only leads to increased thermal comfort but also significantly influences the utilization of the air-conditioning system. It took about 20 minutes with the air-conditioning used in the highest stage to bring the temperature inside the passenger compartment down to a comfortable level when the car with the untreated headliner was driven again. In these twenty minutes, a substantial amount of power was used to provide the desired cooling. On the other side, when the headliner with the phase change material treatment was investigated and the car was driven again, only air-conditioning at the lowest stage was required for about 10 minutes. This leads to substantial energy savings.

10.4.3 Car seat cooling

When the vehicle is parked outside in direct sun exposure, the cover of the driver seat and the front passenger seat heats up. When the driver starts sitting in the seat again and driving the car, the heat stored in the seat is transferred to the driver's body, his skin temperature rises, and he starts to perspire. It takes a while before the heat dissipates from the body, and the seat and comfortable driving conditions are regained. The heat dissipation results from the battery's power-consuming air-conditioning of the passenger compartment [34].

Some seats in upscale vehicles are equipped with seat cooling systems. These systems make seats more thermally comfortable and provide a cooling sensation when someone is sitting in them. These seat cooling installations are either based on forced air provided by a ventilation system or consist of a thermo-electric cooling unit. The operation of both systems requires the utilization of battery power reducing the battery's capacity.

This battery power can be saved when using phase change material for the purpose of seat cooling instead. Patches of a polymer containing phase

change material are arranged at the bottom of the seat. These phase change material patches create a "heat sink", which draws heat away from the seat cover and, more importantly, from the driver's body while staying at a constant temperature as long as the phase change material in them absorbs latent heat. The heat transfer from the seat cover and the driver's body into the seat is accelerated when the body weight compresses the seat cover. The instant heat dissipation ensures a comfortable seat microclimate and, in turn, prevents perspiration by the skin glands.

The seat cooling solution based on the application of phase change material provides a faster response than the ventilation-based and thermo-electric cooling systems. It is very cost-effective, can be accomplished with only minor changes to the seat configuration, and does not require the use of additional equipment. Most importantly, it does not require any external energy supply.

Thermal performance tests

For creating the heat sinks in the seat, a phase change material was selected which starts to absorb latent heat at a temperature of about 28 °C. The phase change material was applied to the test seat in a quantity of about 500 grams, leading to a latent heat storage capacity of about 150 kJ. The microclimate temperature was measured over a one-hour driving period after the vehicle was parked in direct sun exposure for one hour where the seat was unoccupied. The test results are shown in Fig. 10-16.

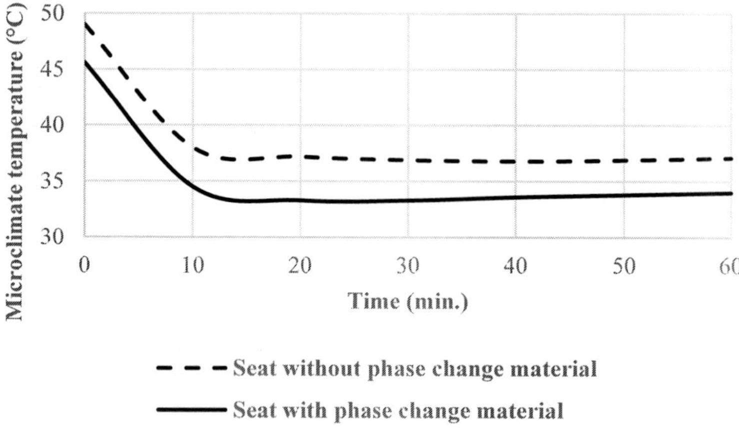

Fig. 10-16: Temperature development in the microclimate.

The phase change material draws heat away from the seat cover before someone sits in it, lowering its temperature at the beginning of the driving period. A substantial amount of heat is transferred from the seat cover to the phase change material inside the seat in the first ten minutes of the driving period.

During driving, the human body constantly produces heat. This heat must be removed from the driver's body to prevent its overheating. Because of the close contact of the driver's body with the seat, part of the generated heat needs to be removed through the seat components. The average amount of heat released into and through the seat components totals about 2 kJ per minute. The heat transfer away from the body is supported by the "heat sink" the phase change material creates. Overall, the initial and continuous transfer of excessive heat away from the body led to a comfortable microclimate temperature of about 33 °C for the remainder of the driving period.

The test of the seat without phase change material shows that the seat cover temperature is high at the beginning of the driving period. When the driver starts sitting in the seat, the heat is transferred from the seat cover to the skin. The heat absorption by the skin will lead to overheating of the body. Because the seat configuration also limits further heat transfer away from the driver's body while driving, the microclimate temperature remains at an uncomfortable 37 °C. The high microclimate, and hence, skin temperature, is the cause of substantial perspiration leading to a high moisture build-up in the microclimate, such as shown in Fig. 10-17. On the other side, the moisture build-up in the microclimate of the seat equipped with phase change material is much lower due to the comfortable microclimate temperature over the entire driving period.

The car seat cooling system does not interfere with a seat heating system because the seat heater is located above the patches equipped with phase change material and the heat generated by the seat heater rises to the top.

The heat stored in the seat, especially in the patches equipped with the phase change material is released into the passenger compartment when the seat is empty again.

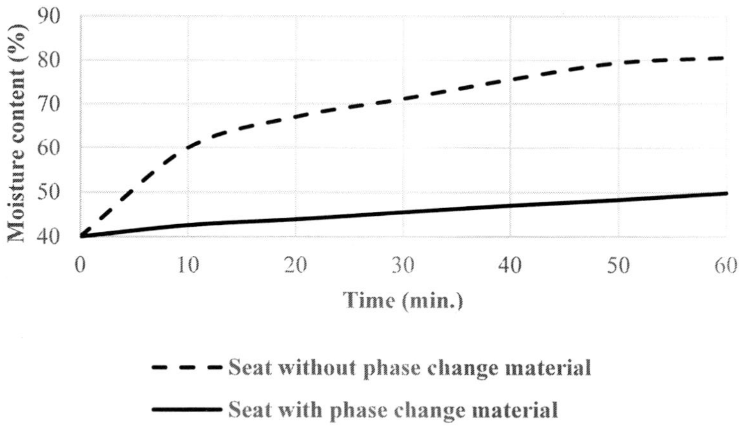

Fig. 10-17: Moisture buildup in the microclimate.

References

1. Cabeza, L.F., Fernandez, C., de Gracia, A., Saffari, M., et al. 2017. "Study on the optimum PCM melting temperature for energy savings in residential buildings worldwide". IOP Conf. Series: Material Science and Engineering. Vol. 251. 012113. DOI: 10.1088/1757-899X/251/1/012113
2. Berardi, Umberto, Gallardo, Andres Alejandro. 2019. "Properties of concretes enhanced with phase change materials for building applications". Energy & Buildings. Vol. 199: 402-414. https://doi.org/10.1016/j.enbuild.2019.07.014
3. Hu, Jong-Wan, Jang, Jeong-Gook, Sharma, Raju. 2022. "Phase-Change Materials in Concrete: Opportunities and Challenges for Sustainable Construction and Building Materials". Materials. Vol. 15, 335. https://doi.org/10.3390/ma15010335
4. Bake, Maitiniyazi, Liu, Shuli, Shukla, Ashish. 2021. "Development of gypsum plasterboard embodied with microencapsulated phase change material for energy efficient buildings". Material Science for Energy Technologies. Vol. 4: 166-176. https://doi.org/10.1016/j.mset.2021.05.001
5. Barreneche, Camila, Cabeza, Luisa F., Fernandez, A. Ines, Haurie, Laia, et al. 2015. "Study of Fresh and Hardening Process Properties of Gypsum with Three Different PCM Inclusion Methods". Materials. Vol.8: 6589-6596. DOI: 10.3390/ma8105324

6. Pause, Barbara. 2001. "Building conditioning technique using phase change materials". Patent No. US 6,230,444
7. Carmona, Manuel, Castilla, Francisco J., Rodriguez, Juan F., Sanz, David, et al. 2018. "Development of Clay Plasters Containing Thermoregulating Microcapsules for Indoor Walls". Intech Open, Sustainable buildings-interaction between a holistic conceptional act and material properties, Chapter 8: 151-167 http://dx.doi.org/10.5772/intechopen.72410
8. Pause, Barbara. 2007. "Wall covering assembly with thermo-regulating properties". Patent No. US 7,241,509
9. Pause, Barbara. 2010. "Taking the heat! Introduction of new energy-efficient building materials based on the application of phase change material (PCM)". International Conference on Building Envelope Systems and Technologies (ICBEST), Vancouver, Canada, June 27-30, 2010
10. Pause, Barbara. 2013. "Enhancing the energy efficiency of buildings by the application of roofing material with phase change material (PCM) treatment". ASTM Symposium, Jacksonville, Florida, USA, October 23, 2013
11. Pause, Barbara Hildegard. 2009. "Building conditioning technique using phase change materials in the roof structure". Patent No. US7,571,758
12. Bernert, Katja. 2019. "Smart wraps". Proceedings of the IASS Annual Symposium 2019 – Structural Membranes 2019, Form &Force: 2379-2391. Barcelona, Spain, October 7-10, 2019
13. Cremers, Jan, Hajek, Oliver, Liebhart, Heiko, Marx, Hannes, et al. 2019. "The ideal ETFE fenestration: The influence of material properties on the thermal performance". Proceedings of the IASS Annual Symposium 2019 – Structural membranes 2019, Form & Force: 1980-1991. Barcelona, Spain, October 7-10, 2019
14. Pause, Barbara Hildegard. 2011. "Membrane materials with thermo-regulating properties for fabric structures". Patent No. US 7,892,988
15. Harvie, G.N. 2015. "Thermal behaviour in architectural fabric structures". In: Fabric structures in architecture. edited by de Llorens, J.I., 221-240, Woodhead Publishing Ltd.
16. Pause, Barbara. 2008. "Taking the heat; New tensile fabrics with thermal-regulating properties: just add phase change material". Fabric Architecture, IFAI publication, Vol. 20, No. 2: 48-51
17. Pause, Barbara. 2013. "New Membrane Materials with Adaptive Heat and Light Transmission Properties". Conference Proceedings Tensinet Symposium: 67-74, Istanbul, Turkey, May 8-10, 2013

18. Pause, Barbara. 2015. "High-performance architectural membranes: phase change material". In: Fabric structures in architecture. edited by de Llorens, J.I., 187-200, Woodhead Publishing Ltd.
19. Campioli, Andrea, Monticelli, Carol, Zanelli, Alessandra. 2013. "Life cycle assessment of textile facades, beyond the current cladding systems". Conference Proceedings Tensinet Symposium: 467-476, Istanbul, Turkey, May 8-10, 2013
20. Pause, Barbara. 2019. "Membrane materials with adaptive heat and light transmission properties for applications in building envelopes". Form & Force Conference, Barcelona, Spain, October 7-10, 2019
21. Heinzelmann, Florian, Teuffel, Patrick. 2013. "Day light performance of multi-layer textile building envelopes". Conference Proceedings Tensinet Symposium: 75-87, Istanbul, Turkey, May 8-10, 2013
22. Bathiebo, Dieudonne Joseph, Guissou, Lamoussa, Ouedraogo, Adama, Ouedraogo, Emmanuel, et al. 2021. "Individual efficiencies of a polycrystalline silicon PV cell versus temperature". Results in Optics, Vol.4. https://doi.org/10.1016/j.rio.2021.100101
23. McCabe, Joseph. 2004. "Optimization of photovoltaic/thermal collectors". Solar Conference Portland, OR, USA. July 11-14, 2004
24. McCormack, Sarah, Griffith, Philip. 2013. "Phase Change Materials: A Primer for Architects & Engineers". European Co-operation in Science and Technology TU0802. www.tu0802.eu
25. Barreneche, Camila, Boer, Dieter, Cabeza, Luisa F., Fernandez, A. Ines, et al. 2018. "Life cycle costing as a bottom line for the life cycle sustainability assessment in the solar energy sector: A review". Solar Energy. https://doi.org/10.1016/j.solener.2018.04.011
26. Al-Abidi, A.A., Anuar Sharif, M.K., Mat. S., Rosli, M.A.M., et al. 2015. "Review of the application of phase change material for heating and domestic hot water systems". Renewable and Sustainable Energy Reviews, Vol. 42: 557-568 http://dx.doi.org/10.1016/j.rser.2014.09.034
27. Pause, Barbara. 2010. "New Textile-Based Hybrid Photovoltaic/Thermal (PV/T) System". World Renewable Energy Congress, Abu Dhabi, UAE, September 25-30, 2010
28. Pause, Barbara. 2019. "New textile-based hybrid photovoltaic/thermal (PV/T) system". 7th International Conference on Intelligent Textiles & Mass Customisation, Marrakesh, Morocco, 13-15 November 2019, IOP Conference Series: Materials Science and Engineering, Vol. 827. 2020 https://iopscience.iop.org/issue/1757-899X/827/1

29. Pause, Barbara Hildegard. 2011. “Textile heat accumulator for utilization of solar energy”. Patent No. US 7,913,685
30. Barreneche, Camila, Ding, Yulong, Jin, Lu, Lin, Cheng, et al.2022. “Thermal energy storage for electric vehicles at low temperature: Concepts, systems, devices and materials”. Renewable and Sustainable Energy Reviews. Vol. 160. https://doi.org/10.1016/j.rser.2022.112263
31. Biglarbegian, M., Gadsden, S.A., Lyu, Y., Mahmud, S., et al. 2019. “Electric vehicle battery thermal management system with thermoelectric cooling”. Energy Reports. Vol. 5: 822-827 https://doi.org/10.1016/j.egyr.2019.06.016
32. Pause, Barbara. 2019. “Thermal control of automotive systems by the application of macro-encapsulated phase change material”. Thermal Materials Summit, Los Angeles, May 1-2, 2019
33. Brütting, Michael, Haussmann, Thomas, Mehling, Harald. 2022. “PCM products and their fields of application – An overview of the state in 2020/2021”. Journal of Energy Storage, 51. https://doi.org/10.1016/j.est.2022.104354
34. Pause, Barbara. 2016. “Taking the heat – thermal control of the automotive interior with phase change material (PCM)”. IFAI Expo, Charlotte, USA. October 18-19, 2016
35. Pause, Barbara Hildegard. 2008. “Thermal control of automotive interiors with phase change material”. Patent No. US 7,320,357

CHAPTER ELEVEN

SUMMARY OF THE DESIGN GUIDELINES

For a successful design and development of an end-use product equipped with phase change material the following steps are recommended to be carried out:

1. Tests of the original product to determine application conditions as well as the heat transfer characteristics under the prevailing application conditions.
2. Predetermination of the possible location of the phase change material in the end-use product.
3. Calculation of the temperature range in which the phase change material will presumably operate at this location based on the heat transfer characteristics of the product components.
4. Pre-selection of phase change material candidates that operate in the calculated temperature range.
5. Estimating the latent heat storage capacity necessary to obtain expected thermal effects.
6. Calculation of the required quantity of each phase change material candidate based on its latent heat storage capacity.
7. Narrowing the number of phase change material candidates based on requirements regarding properties such as non-combustible, non-corrosive, non-toxic, as well as sufficient thermocycling behavior, etc.
8. Limitations in weight addition and available space required by the application should be considered to narrow down possible phase change material candidates.
9. Further elimination of possible candidates based on prices and availability of the preselected phase change materials.
10. Determination of whether encapsulation is required.
11. Selection of an appropriate form of containment such as fibers, coatings, films, etc.
12. Selection of suitable carrier materials, if necessary.
13. Determination if thermal conductivity enhancing materials or nucleation agents need to be applied.

14. Determination of the final location of the phase change material in the end-use product.
15. Manufacturing of the substrate with the phase change material treatment.
16. Manufacturing of a prototype of the end-use product.
17. Tests of the thermal performance of the prototype of the end-use product with the phase change material application compared to the original product.
18. Quantification of the thermal performance enhancement by the phase change material application.
19. Determination of production procedures as well as sales and marketing strategies.

INDEX

S

T

V

W